MAX-PLANCK-INSTITUT
FÜR AUSLÄNDISCHES ÖFFENTLICHES RECHT
UND VÖLKERRECHT

Beiträge zum ausländischen öffentlichen Recht und Völkerrecht

Begründet von Viktor Bruns

Herausgegeben von
Armin von Bogdandy • Anne Peters

Band 290

Anne Peters
Editor

Studies in Global Animal Law

 Springer Open

ISSN 0172-4770   ISSN 2197-7135   (electronic)
Beiträge zum ausländischen öffentlichen Recht und Völkerrecht
ISBN 978-3-662-60755-8   ISBN 978-3-662-60756-5   (eBook)
https://doi.org/10.1007/978-3-662-60756-5

This book is an open access publication.

This Springer imprint is published by the registered company Springer-Verlag GmbH, DE, part of Springer Nature.
The registered company address is: Heidelberger Platz 3, 14197 Berlin, Germany

Acknowledgments

This book had a long way to come to fruition. Several chapters build on prior online-only publications; others (the introductory chapter and Chapters 9 and 13) have been written completely fresh. Chapters 2–8, 11–12, and 14 are updated and mostly expanded versions of essays which were first published online in the *American Journal of International Law Unbound* 111 (2017), Symposium on Global Animal Law (pp. 252–281 and 397–424). Chapter 10 is an expanded and revised version of an essay in the *American Journal of International Law Unbound* 112 (2018), 355–360.

I gratefully acknowledge the permission of the editors and the managing editor (then Melissa Durkee) of AJIL Unbound to use those essays for this book. I also thank Brigitte Reschke from the publishing house Springer for offering the opportunity to present the work in a consolidated form. I am indebted to my students of the master's programme 'Animal Law and Society' at the Autonomous University of Barcelona, successfully run by professor Marita Candela, now in its ninth edition, for critical questions and their enthusiasm in developing global animal law. Last but not least, I thank Anette Kreutzfeld for her precious editorial and logistic support. The book *Studies in Global Animal Law* is dedicated to the memory of Felix and Cora.

Acknowledgments

Contents

Part III   New Legal Concepts

Part IV   New Protective Legal Strategies

Chapter 1
Introduction

Anne Peters

Abstract The introduction explains key concepts and methods. It defines global animal law as the sum of legal rules and principles governing the interactions between humans and other animals, on a domestic, local, regional, and international level. Global animal law reacts to the mismatch between almost exclusively national animal-related legislation on the one hand, and the global dimension of the animal issue on the other hand. The merely national regulation of animal welfare within the states' boundaries runs aloof in the face of globalisation. This gives rise to an animal welfare gap. Moreover, animal use creates global problems ranging from climate and soil degradation over antimicrobial resistance to food insecurity. This requires a global law response. The introduction also gives a brief overview over the book and its main findings.

1 Global Animal Law in a Nutshell

The essays assembled in this volume analyse both foundational and current legal aspects of human—animal relationships from a global animal law perspective.

Global animal law refers to the sum of legal rules and principles (both state made and non-state made) governing the interaction between humans and other animals, on a domestic, local, regional, and international level. Given that the various 'levels' of regulation are normally not neatly separate but rather intermesh and criss-cross, the image of 'marbled' regulation might be preferable to 'multi-layered'. The body of global animal law[1] comprises hard law in the form of statutes and treaties and soft law such as standards issued by international organisations and voluntary labelling

---

[1] An alternative label would be 'transnational animal law'. However, the term 'global' expresses better that some relevant problems are currently addressed only in national law. In addition, the

A. Peters (✉)
Max Planck Institute for Comparative Public Law and International Law, Heidelberg, Germany
e-mail: apeters-office@mpil.de

© The Author(s) 2020
A. Peters (ed.), *Studies in Global Animal Law*, Beiträge zum ausländischen öffentlichen Recht und Völkerrecht 290,
https://doi.org/10.1007/978-3-662-60756-5_1

schemes or industry norms and codes of conduct. So the actual producers of global animal law are parliaments, governments, international institutions, business, civil society organisations, the latter often acting transnationally and in collaboration with governmental agencies. The scholarly analysis and commentary on these bodies of law constitute the academic discipline of global animal legal studies, or—for simplicity's sake—the discipline of global animal law.

Global animal law comprises but significantly moves beyond the international legal instruments which seek to conserve endangered species,[2] to protect wild animal habitats,[3] and to uphold biological diversity.[4] These rules pay attention to collective goods, mainly for anthropocentric reasons. In contrast, they barely address the welfare of individual animals or potential rights of some animals which results in the notorious wild animal welfare gap.[5] Global animal law also seeks to emancipate itself from classic environmental law and bears some overlap with the novel branch of biolaw.[6]

Speaking of 'global animal law' conveys the message that animal law can be best understood, applied and analysed when legal practitioners and scholars have an eye simultaneously on the various layers of the law, and on various norm types. The corpus of domestic, international, and local law, of state-made and privately generated, of hard and soft law relating to the treatment and welfare of animals has reached a critical mass which justifies summing it up as a cross-cutting matter or as a legal field of its own, under one overarching heading.

2   A Globalised Problem Requires a Global Solution

Global animal law is the response to the mismatch between almost exclusively national animal-related legislation on the one hand, and the global dimension of the animal issue on the other hand. States increasingly regulate animals unilaterally

---

qualifier 'global' in a wide sense conveys that the approach is universalist, comprehensive, and holistic. See Brels, 'Global Approach' 2017, 105.

[2]Such as the Convention on International Trade in Endangered Species of Wild Fauna and Flora (CITES), 3 March 1973, 993 UNTS 243 (entered into force 1 July 1975) and the Bonn Convention on the Conservation of Migratory Species of Wild Animals (CMS), 23 June 1979, 1651 UNTS 333 (entered into force 1 November 1983).

[3]See, e.g., Council of the European Communities, Council Directive 92/43/EEC on the Conservation of Natural Habitats and of Wild Fauna and Flora, 21 May 1992, Official Journal EU L 206, 22 July 1992, 7-50.

[4]See, e.g., Convention on Biological Diversity (CBD), 5 June 1992, 1760 UNTS 79 (entered into force 29 December 1993).

[5]See for a recent scholarly collection seeking to expand and intensify the international law approach: L'Observateur des Nations Unies: Revue de l'Association française pour les Nations Unies 45 (2018), 'L'animal'.

[6]See, e.g., Carporale/Pavone, International biolaw 2018.

through animal welfare and protection laws, but the limited scope of national regulation within the states' confines hampers their effectiveness. They run aloof in the face of globalisation. Because animal issues have 'gone global', they require global responses of the law, ideally in combination with local solutions.[7]

Why and how have animal issues gone global? A number of drivers and manifestations should be mentioned here. First, virtually all aspects of (commodified) human−animal interactions (ranging from food production and distribution, working animals, animal use in research, to breeding and keeping of pets) possess a transboundary, a global, dimension. The industrialisation of meat, dairy, and fur production has massive environmental, climatic, social, and ethical consequences on a global scale. For example, health costs ascribed to the excessive intake of animal-based food arise everywhere in the world.[8] Global warming is induced, inter alia, by the abundance of cattle waste.[9] Antimicrobial resistance triggered mainly by the excessive use of antibiotics in high-density industrial farming is a global problem for human health. The loss of genetic information through the extinction of species concerns all mankind. Armed conflict in Africa is financed by wildlife poaching which is sustained by global criminal networks and illegal markets spanning to Asia.[10] Piracy off the African coasts is fuelled by the loss of income of local fishermen due to the global overexploitation of fish stocks and poses a global traffic problem. In all these examples, what is at issue are sustainability, the extinction of species, poverty, and malnutrition—all of which are global problems.

Second, growing consumer attention to animal welfare aspects in their purchasing decisions on products involving or using animals has an impact both on the industries and governments. Consumers in industrialised countries increasingly expect law-makers (for varying reasons, including anthropocentric ones such as human health and fitness) to take animal welfare seriously. The resulting political pressure not only affects the regulation of domestic production but also that of the importation of foreign animal products. For example, according to EU-wide polls, 93% of Europeans agree that 'imported products from outside the EU should respect the same animal welfare standards as those applied in the EU'.[11] 'Animal-friendly' states or actors such as the EU therefore tend to either 'export' their animal welfare standards by demanding certifications on identical or equivalent production methods

---

[7]Park/Singer, 'Globalization of Animal Welfare', 2012; Peters, 'Global Animal Law' 2016; Brels, 'Global Approach' 2017.

[8]Willett et al., 'Food in the Anthropocene' 2019.

[9]Food and Agriculture Organisation (FAO), Livestock's Long Shadow, Rome 2006, 271.

[10]See United Nations, General Assembly, Resolution 71/326 'Tackling illicit trafficking in wildlife', UN Doc. A/RES/71/326, 11 September 2017; United Nations, Security Council, Resolution 2399 (2018), UN Doc. S/RES/2399 (2018), 30 January 2018, preamble 18th indent.

[11]Survey requested by the European Commission, Directorate-General for Health and Food Safety, Special Eurobarometer 442: Attitudes of Europeans towards Animal Welfare, March 2016, 27-28.

(e.g. slaughter[12]), require labelling to prove equivalent production methods (e.g. cage-free eggs), or ban imports altogether. A prominent example for the latter is the EU's prohibition of the import of seal products which the EU chiefly justified 'in response to public moral concerns about the animal welfare aspects of the killing of seals and the possible presence on the Union market of products obtained from seals killed in a way that causes excessive pain, distress, fear and other forms of suffering'.[13] Such rising consumer awareness is a stimulus for regulatory response, which needs to be global against the background of global trade.

Third, and relatedly, businesses seeking to export their animal products into states in which the consumers are more attentive to animal welfare are paying more attention to the issue because they do not want to lose market shares. This also holds for the regulators in the countries of export if they want to support their trading industries. Along this line, a FAO report noted: 'Animal welfare is not a new subject for regulation in most developed countries, owing to a sophisticated consumer base and greater exposure to animal welfare issues. Growing international trade is generating more interest in animal welfare elsewhere in the world, in particular in countries seeking to increase trade with Europe.'[14] In short, market players in all regions of the world demand a harmonised regulation of animal welfare and animal protection.

A fourth, pragmatic reason for developing global animal welfare standards is that these can provide a benchmark for local, national, and international legislation. At present, animal welfare or rights activists face the daunting and repetitive task of battling for new laws in multiple, isolated national jurisdictions. An international yardstick would allow them to devote their scarce resources on implementation of that acknowledged standard.[15]

The fifth impetus for global animal law is the need for interpretative guidance. Various international instruments, notably trade agreements, directly or indirectly concern animals. They must be applied and to this end interpreted. New international rules on animal welfare could serve at this point. The pertinent prescription of the Vienna Convention on the Law of Treaties (VCLT) requires that 'there shall be taken into account (...) any relevant *rules of international law* applicable in the relations between the parties' (Art. 31(3)(c) VCLT).[16] Only international, not domestic, rules

---

[12]Art. 12(1) of Council Regulation EC 1099/2009 on the protection of animals at the time of killing, 24 September 2009, Official Journal EU L 303/1, 18 November 2009.

[13]Regulation (EU) No. 2015/1775 of the European Parliament and of the Council of 6 October 2015 amending Regulation (EC) No. 1007/2009 on trade in seal products and repealing Commission Regulation (EU) No. 737/2010, OJ 2015, L 262/1, consideration 1. That regulation was amended in order to comply with a WTO decision. The amended regulation was Regulation (EC) No. 1007/2009 on Trade in Seal Products, 2009, OJ EU L 286/36. Its cons. 1 stated: 'Seals are sentient beings that can experience pain, distress, fear and other forms of suffering. (...)'.

[14]Kuemlangan, 'Preface' 2010, V.

[15]Favre, 'International Treaty' 2012, 239.

[16]The precondition of art. 31 VCLT and its underlying principle, namely that the rule be 'applicable to the relations between the parties', has been construed infamously narrowly by the WTO *Biotech*

could perform this important function as a catalyst for a more animal-friendly reading of existing international trade agreements.

These observations, taken alone, justify a global law approach to animal welfare. The need for such an approach becomes dramatic when we consider the most important features of globalisation, namely capital and labour mobility and global supply chains in the animal industries, to which we now turn.

3 Global Animal Law as a Response to Outsourcing

One of the most important motivations for Global animal law is the urgent need to close the legal loopholes available to animal-related industries exploiting the opportunity to migrate away from stringent national animal welfare standards. The animal-processing industry (for food and pharmaceuticals) is a global industry and thrives on global trade. Even if one country attempts to improve welfare standards, for example for the caging of livestock, for slaughter, or for animal experiments, it cannot do so unilaterally if it wants to be effective. The reason is that the affected sectors or branches of industry can escape stricter regulations by relocating.[17] Such relocation, or 'leakage', to cheap and low-standard countries then renders high national animal protection standards meaningless. A concrete example for such evasion is the transfer of the slaughter of horses from the United States (US) to Mexico, where welfare standards are much lower. After closure of the last horse slaughter facilities in the US in 2007, exportation of horses for slaughter to Canada and Mexico has increased dramatically (by 660% to Mexico), with unintended negative effects on horses' welfare, notably during transport.[18] Another pertinent field is biomedical research. In 2010, an international group of researchers, research funding institutions, and representatives of pharmaceutical and biomedical industries

---

panel. The panel noted that the Cartagena Protocol, on which the European Community as a respondent had relied for interpreting the pertinent WTO Agreements, was in fact 'not applicable', because the Protocol had not been ratified by a number of WTO members, including the complaining parties to the dispute (USA, Argentina, and Canada). World Trade Organization, European Communities – Measures Affecting the Approval and Marketing of Biotech Products, WT/DS 291–293/R, 29 June 2006, paras. 7.49 – 7.95, notably para. 7.75. The WTO Appellate Body in the *Airbus* case moved away from the *Biotech* approach (World Trade Organization, Appellate Body, European Communities and certain Member States – Measures affecting Trade in Large Civil Aircraft, WT/DS 316/AB/R, 18 May 2011, paras. 839-855). The better and now prevailing view is that it is not necessary that all states in the organisation/treaty are also parties to the other treaty to make the latter usable, if they are not involved in the dispute.

[17]Peters, 'Competition between Legal Orders' 2014.

[18]US Government Accountability Office, Report to Congressional Committee, 'Horse Welfare: Action Needed to Address Unintended Consequences from Cessation of Domestic Slaughter', GAO-11-228, 22 June 2011.

adopted the Basel Declaration,[19] with the (implicit) objective of persuading regulators and the public to renounce overly strict regulation of biomedical research using
animals. The accompanying statements underscored the importance of preserving
Switzerland as a site for biomedical research if research regulation became too strict.
The organisers at least implicitly raised the danger of outsourcing the industry,
which in turn would lead to important losses of tax income for Switzerland. A
final example is a statement of the German Research Foundation (*Deutsche
Forschungsgemeinschaft*) that complained about overly bureaucratic procedures
for obtaining permits to perform animal experiments in Germany.[20] Such complaints
might have a chilling effect on regulators. Moreover, there is already a trend that
researchers seek to evade strict standards through new forms of research collaborations with colleagues in low-standard countries, or fully move to those states.[21] If
regulators bow to such pressures, and when individual countries try to keep or regain
economically significant industrial sectors by supplying an attractively permissive
legal environment, the further elevation of standards is stalled (frozen), and in the
worst case a downward spiral of standards could be set in motion, a race to the
bottom, to the detriment of the welfare of animals.[22]

Nation states with a high level of high animal protection, conscious of the global
competition among states over mobile industries, face various policy options.[23] The
obvious response might be to lower national standards (resulting in the
abovementioned 'race to the bottom'). An alternative path is to campaign for
uniform international rules in order to prevent other states from exploiting their
lower or lacking requirements as an (unfair) competitive advantage. Uniform international law can level the playing field for their firms from high-standard states by
subjecting all businesses to one and the same (high) norm.[24] This strategy is
employed with regard to animal welfare standards by the EU which has relatively

---

[19]Adopted on the occasion of the first Basel conference 'Research at a Crossroads', Basel (Switzerland), 29 November 2010, available at: http://www.basel-declaration.org. The official wording is
a 'call for more trust, transparency and communication on animal research'.

[20]Deutsche Forschungsgemeinschaft, Ständige Senatskommission für tierexperimentelle
Forschung, 'Genehmigungsverfahren für Tierversuche: Stellungnahme der Ständigen
Senatskommission für tierexperimentelle Forschung der Deutschen Forschungsgemeinschaft
(DFG)', 5 September 2018, available at: http://www.dfg.de/service/presse/pressemitteilungen/
2018/pressemitteilung_nr_37/index.html.

[21]Sueur, 'La fuite de la recherche biomédicale' 2016, 19; Chatfield/Morton, 'Use of Non-human
Primates' 2018, 81-90 on researchers' evasion to low standard countries through collaborative
ventures.

[22]See also Anne Peters, Chap. 10 in this volume, Charlotte Blattner, Chap. 12 in this volume.

[23]See for a reform proposal for raising welfare standards for farmed animals in Germany that takes
into account the danger of a migration of the industry beyond the state boundaries, together with a
number of suggestions about how to prevent such migration: Bundesministerium für Ernährung und
Landwirtschaft, Wissenschaftlicher Beirat für Agrarpolitik, 'Wege zu einer gesellschaftlich
akzeptierten Nutztierhaltung: Gutachten', Berlin, March 2015.

[24]Murphy, *Regulatory Competition* 2004; Baldwin/Cave/Lodge, *Understanding Regulation* 2012,
chapter 17 'Regulatory Competition and Coordination', 356-369.

more stringent animal welfare laws than most of its global trading partners. The EU Commission solicited a study on the impact of animal welfare rules on the international competitiveness of agricultural operators outside of the EU. In its 2018 report on that study, the EU Commission stated that '[t]he overall objective of the Commission's international activities on animal welfare is promoting EU values regarding animals, (...) and encourage[ing] globally, particularly with EU-trading partners, high animal welfare standards, *reflecting the EU model and principles.* Improving animal welfare standards globally also contributes to ensure a *level playing field* between EU and non-EU operators. (...) to be sustainable, [a legislative model on animal welfare] *should also be disseminated internationally'.*[25] In the early years of the WTO, the EU had proposed that the trade organisation should directly address animal welfare standards.[26] The EU's motivation is to avert 'that its animal welfare standards could be undermined and that it could suffer negative trade effects, since agricultural products produced to meet high EU animal welfare standards would run the risk of being edged out of the market by cheaper imports produced under lower standards'.[27] The proposal for adopting a WTO-wide animal welfare standard thus far has not found favour with WTO members but is maybe anyway moot due to the current paralysis of the WTO.

Since then, at least ten new bilateral and regional "deep" trade agreements foresee the exchange of information, 'dialogue', 'consultation', and cooperation, collaboration, and/or technical assistance on animal welfare (Annex IV, Art. 12(2)(e) of the Agreement EU—Chile (2002),[28] Art. 5.1. sec. 2 and Art. 5.9 of the Agreement EU—South Korea (2010),[29] Art. 62 of the Association Agreement EU—Central American States (2012; trade part provisionally applied since 2013),[30] Art. 102 of the Agreement EU—Andean states (Colombia, Peru, and Ecuador, potentially Bolivia)

---

[25]European Commission, Report from the Commission to the European Parliament and the Council on the impact of animal welfare international activities on the competitiveness of European livestock producers in a globalised world, COM(2018) 42 final, 26 January 2018, 1 (emphases added).

[26]World Trade Organization, Committee on Agriculture, European Communities Proposal: Animal Welfare and Trade in Agriculture, G/AG/NG/W/19, 28 June 2000.

[27]Vapnek/Chapman, *FAO*, 17.

[28]Agreement establishing an association between the European Community and its Member States, of the one part, and the Republic of Chile, of the other part, signed on 18 November 2002, entered into force on 1 March 2005 (OJ EU 2002 L 352, 3). See Annex IV, art. 1(2): 'This Agreement aims at reaching a common understanding between the Parties concerning animal welfare standards.'

[29]Free Trade Agreement between the European Union and its Member States, of the one part, and the Republic of Korea, of the other part of 6 October 2010 (OJ EU L 2011 127, 1).

[30]Agreement establishing an Association between the European Union and its Member States, on the one hand, and Central America on the other (OJ EU 2012 L 346, 3). The Central American state parties are Costa Rica, El Salvador, Gutatemala, Honduras, Nicaragua, Panama. The agreement is not yet in force but the trade part is provisionally applied since 2013.

(2012),[31] Sec. 2 of the EU—Brazil Memorandum of Understanding (2013),[32] Art. 68(4) and Art. 404 of the Association Agreement EU—Ukraine (2014),[33] Art. 59 (4) of the Association Agreement EU—Georgia (2014),[34] Art. 21(4) lit. s) of CETA (2016),[35] Art. 35 of the Agreement EU—Philippines (2017),[36] Art. 18.17 of the Agreement EU—Japan (2017),[37] Art. 16.3 of the Agreement E—Vietnam (2018)).[38] In 2018, an agreement of principle on the modernisation of the EU−Mexico Global Agreement was reached which foresees an entire chapter on 'Cooperation in Animal Welfare and Anti-Microbial Resistance'.[39] The mentioned cooperation and capacity-building provisions are placed either in an SPS-'plus' chapter or in a separate chapter on regulatory cooperation.

To conclude, a downwards spiral of animal welfare and protection standards can be prevented only by the dissemination of adequate standards worldwide, and we have seen that such dissemination strategies are already underway, promoted notably by the EU.

---

[31]Trade Agreement between the European Union and its Member States, of the one part, and Colombia and Peru, of the other part (OJ EU 2012 L 354, 3). The agreement is provisionally applied since 2013 to Colombia and Peru, since 2017 also provisionally applied to Ecuador.

[32]Administrative Memorandum of Understanding on Technical Cooperation in the Area of Animal Welfare between the Ministry of Agriculture, Livestock and Food Supply of the Federative Republic of Brazil and the Directorate General of Health and Consumers of the European Commission, 24/01/2013, available at http://www.itamaraty.gov.br/en/press-releases/16365-acts-signed-on-occasion-of-the-6th-brazil-european-union-summit-brasilia-january-24#2agreem.

[33]Association Agreement between the European Union and its Member States, of the one part, and Ukraine, of the other part (OJ EU 2014 L 161, 3).

[34]Association Agreement between the European Union and the European Atomic Energy Community and their Member States, of the one part, and Georgia, of the other part (OJ EU 2014 L 261, 4).

[35]Comprehensive Economic and Trade Agreement (CETA) between Canada, of the one part, and the European Union and its Member States of the other part of 30 October 2016, ratified by EU on 15 February 2017, provisional entry into force on 21 September 2017 (OJ EU 2017 L 11, 23).

[36]European Commission, 'EU Textual Proposal – EU-Philippines Free Trade Agreement: sanitary and phytosanitary measures' (January 2017), available at: http://trade.ec.europa.eu/doclib/docs/2017/march/tradoc_155432.pdf.

[37]Agreement between the European Union and Japan for an Economic Partnership of 17 July 2017, OJ EU 2018 L 330, 4.

[38]Text as of August 2018; provisionally applied, available at: http://trade.ec.europa.eu/doclib/press/index.cfm?id=1437.

[39]The EU Commission published the texts of the Trade Part of the Agreement following the agreement in principle announced on 21 April 2018. The provisional text is: '1. The Parties recognise that animals are sentient beings. 2. The Parties recognise the value of the OIE animal welfare standards, and shall endeavour to improve their implementation while respecting their right to determine the level of their science-based measures on the basis of OIE animal welfare standards. 3. The Parties undertake to cooperate in international fora with the aim to promote the further development of good animal welfare practices and their implementation. The Parties recognise the value of increased research collaboration in the area of animal welfare.'

4 Global Animal Law as an Analytical Lens

Global animal law is first of all critical—more deeply critical than classical animal welfare law, animal protection law, and wildlife law. It moves away from merely lamenting the weak legal protection for animals and suggesting reforms. Rather, it starts from the insight that the law is profoundly ambivalent in its approach to animals: it not only serves to protect animals from individual deviant abusive behaviour but also perpetuates institutional violence against animals.[40]

One policy claim of global legal animal studies is that legal rules for the benefit of animals, their status, their welfare and potential rights can be effective only if they are enacted both on the domestic and on the international level. Obviously, the regulatory response must grow from the bottom up. International law-making institutions have no chance of imposing rules on states that do not take sufficient cognizance of animal issues in their own domestic law. A domestic legal basis must form the breeding-ground for international norms and must secure their operation.

A follow-up question is how much in this area can be left to so-called indirect regulation through the invisible hand of the market, and where 'command-and-control' regulation is needed. The 'market-based approach' is less paternalistic and a good compromise for governments whose citizenship is divided about animal welfare. If only a minority is highly critical of a given animal production method, let them decide for themselves with their purse to avoid participating in these practices as consumers. However, and as a matter of principle, the meeting of offer and demand on a market can bring about appropriate product and production standards only when consumers are comprehensively informed before making their purchasing choices. This is typically lacking in the animal-related industry. This means that the first level of regulation by the states should aim at transparency, consumer information, certification, and labelling. Only on this basis, market-based regulation can function at all. Another drawback of the purely market-based approach is that it favours those who are willing and able to consume over those who are unable or unwilling to consume, or both. Vegetarians cannot vote with their chequebook on animal-friendly meat production. Also, such 'regulation' is less effective because it is less complete; a more or less large residual market for the unwanted animal product will almost always persist. In result, animal welfare and protective regulation will need a combination of state made and non-state rules.

Finally, global animal legal scholars actively embrace the new approaches in ethics, political theory, and social anthropology that have generated the fields of human—animal studies (HAS),[41] animal politics,[42] and critical animal theory[43]—

---

[40]Seminally Caspar, *Tierschutz im Recht der modernen Industriegesellschaft* 1999; Bolliger, *Europäisches Tierrecht* 2000. More recently Michel/Kühne/Hänni, *Animal Law* 2012.

[41]Marvin/McHugh, *Human—Animal Studies* 2014.

[42]Seminally Donaldson/Kymlicka, *Zoopolis* 2011. See also Pelluchon, *Manifeste animaliste* 2016.

[43]See *The Journal for Critical Animal Studies* (since 2003; http://journalforcriticalanimalstudies.org/).

trends celebrated as constituting an 'animal turn'[44] in the social sciences and humanities.

In conclusion, global animal law functions as an umbrella term that allows us to grasp the complex nature and characteristics of the pertinent legal issues better, and thus to better analyse, criticise, and advance the legal regimes governing animals globally. Each chapter of this book seeks to make a distinct contribution to this end.

5 The Contributions to this Volume

The contributions delve into the history of the *ius gentium,* examine various aspects of how national and international law traditionally deals with animals as commodity, and finally suggest new legal concepts and protective strategies.

Part One lays historical foundations. Two chapters, written by historians, demonstrate that scholars of the *ius naturae et gentium* from the sixteenth to the seventeenth centuries contemplated whether and how to include animals in the sphere of politics and justice. The *ius naturae* was premised on an idea of human nature, and this idea was developed partly in contradistinction to animal nature.

Annabel Brett (Chap. 2) shows that animals were not totally excluded from any kind of right and that violence against them was not always regarded as legitimate. In Chap. 3, Anna Becker traces how early modern writers of political theory, often in their comments on Aristotle, viewed the relationships between some animals and humans, notably in the household.

In Chap. 4, Mathilde Cohen examines 'animal colonialism'. European conquerors and settlers exported the technique of dairy production to all parts of the world. By propagating and spreading animal milk consumption and depreciating colonised women's practice of breastfeeding, the oppression of humans and animals went hand in hand.

Part Two deals with animals as commodity. Chapter 5 by Kristen Stilt examines the trade of live animals for slaughter, focusing on export from Australia to the Muslim-majority countries that are the main customers. The current legal regime governing live exports is insufficient to provide animals with an adequate standard of welfare. But with the due involvement of religious authorities, the Islamic tradition of animal welfare could be harnessed to develop more widely accepted international transportation and slaughtering standards.

Chapter 6 by Stefan Kirchner discusses animal use by indigenous peoples that involve crossing state borders, using the example of reindeer herding by Sámi in Sweden, Norway, and Finland. Animals play important cultural, economic, and spiritual roles for indigenous communities which are not sufficiently recognised by contemporary laws. The risk of overruling the interests of migratory animals and of

---

[44]Ritvo, 'On the Animal Turn' 2007, 118-122.

the pastoralist (semi-)nomadic human communities depending on them, is exacerbated when the herds cross boundaries.

Chapter 7 by Jiwen Chang gives an account of China's new legal framework (particularly the Wild Animal Protection Law of 2016). Chang suggests several concrete measures for improvement, including the introduction of public interest litigation, better coordination among governmental departments, a trading information platform, and consultation with the secretariat of the Convention on International Trade in Endangered Species of Wild Flora and Fauna (CITES), in order to bring the Chinese legal and administrative framework fully in line with CITES.

In Chap. 8, Radha Ivory sketches how the international topics of corruption and endangered animal trafficking have been connected in hard and soft international law, including by United Nations Security Council resolutions. Ivory cautions against linking the two legal frameworks and reform agendas, inter alia because the combined anti-corruption/wildlife trafficking discourse may distract from the vigilance against illicit investment and excessive consumption in the Global North, which enable and drive the crimes.

Part Three introduces new legal concepts. In Chap. 9, Guillaume Futhazar explores the place of the concept of animal welfare in biodiversity and species protection agreements. He suggests that new international rules aiming at ensuring the protection of wild animals' welfare could fulfil a double purpose: strengthening conservation and filling the welfare gap of international biodiversity law.

In Chap. 10, Anne Peters argues that animal rights could and should be recognised by international law. Animal rights would complement human rights not the least because the entrenchment of the species-hierarchy as manifest in the denial of animal rights in the extreme case condones disrespect for the rights of humans themselves.

In Chap. 11, Saskia Stucki examines the labelling of animal products as 'humane' and likens the idea of humanising animal slaughter, factory farms, and other forms of production to the notion of humanising warfare. Like international humanitarian law, animal welfare law is marked by the tension inherent in its attempt to humanise innately inhumane practices. Both areas of law endorse a principle of 'humanity' while arguably facilitating and legitimising the use of violence, and might thereby ultimately perpetuate the suffering of living beings.

Part Four explores new protective legal strategies. In Chap. 12, Charlotte Blattner examines how extraterritorial jurisdiction can help to overcome regulatory gaps in animal law, much as criminal law or antitrust law successfully responded to global problems through laws that reach across borders. Because the emergence of an international treaty regulating animal abuse is currently unlikely, extraterritorial animal law, if applied reasonably, could fundamentally improve the protection of animals, both those located at home and abroad.

In Chap. 13, Tom Sparks discusses the potential of a human rights framework to contribute to the growth and development of global animal law, taking as example the jurisprudence of the European Court of Human Rights. Sparks concludes that although the telos of human rights law is different from that of animal law,

nevertheless there exist many overlapping concerns within which mutually beneficial interactions are possible.

Chapter 14 by Jérôme de Hemptinne turns to the treatment of animals in international humanitarian law (IHL). IHL does not contain explicit rules to mitigate the
suffering of animals in armed conflict. However, the overall evolution of law's
approach to animals, notably its recognition of them as sentient beings, appears to
allow for a progressive interpretation of IHL so as to constrain acts of violence
against animals in war.

As this book demonstrates, legal scholars concerned with animal issues are
developing proposals to fill gaps in international law, are reformulating traditional
legal concepts such as rights, jurisdiction, or civilians, and are reconfiguring the
domestic law–international law divide. By showing numerous entry points for
animal issues in international law and at the same time shifting the focus and
scope of inquiry, the book seeks to push forward the field of global animal law
and global legal animal studies as a scholarly discipline.

References

Baldwin, R., Cave, M., & Lodge, M. (2012). *Understanding regulation: Theory, strategy, and practice* (2nd ed.). New York: Oxford University Press.

Bolliger, G. (2000). *Europäisches Tierrecht*. Zürich: Schulthess.

Brels, S. (2017). A global approach to animal protection. *Journal of International Wildlife Law & Policy, 20*, 105–123.

Carporale, C., & Pavone, I. (Eds.). (2018). *International biolaw and shared ethical principles: The universal declaration on bioethics and human rights*. London: Routledge.

Caspar, J. (1999). *Tierschutz im Recht der modernen Industriegesellschaft: Eine rechtliche Neukonstruktion auf philosophischer und historischer Grundlage*. Baden-Baden: Nomos.

Chatfield, K., & Morton, D. (2018). The use of non-human primates in research. In D. Schroeder, J. Cook, F. Hirsch, S. Fenet, & V. Muthuswamy (Eds.), *Ethics dumping. Case studies from North-South research collaborations* (pp. 81–90). Cham: Springer.

Donaldson, S., & Kymlicka, W. (2011). *Zoopolis: A political theory of animal rights*. Oxford: Oxford University Press.

Favre, D. (2012). An international treaty for animal welfare. *Animal Law Review, 18*, 237–279.

Kuemlangan, B. (2010). Preface. In J. Vapnek & M. Chapman (Eds.), *Legislative and regulatory options for animal welfare, for the development law service*, FAO Legal Office. FAO Legislative Study 104. Rome: Food and Agriculture Organization of the United Nations.

L'observateur des Nations Unies: Revue de l'Association française pour les Nations Unies, *45*(2) (2018) 'L'animal'.

Marvin, G., & McHugh, S. (Eds.). (2014). *Routledge handbook of human-animal studies*. New York: Routledge.

Michel, M., Kühne, D., & Hänni, J. (Eds.). (2012). *Animal law – Tier und Recht: Developments and perspectives in the 21st century*. Zürich: Dike.

Murphy, D. D. (2004). *The structure of regulatory competition*. Oxford: Oxford University Press.

Park, M., & Singer, P. (2012). The globalization of animal welfare: More food does not require more suffering. *Foreign Affairs, 91*, 122–133.

Pelluchon, C. (2016). *Manifeste animaliste: Politiser la cause animale*. Paris: Alma éditeur.

Peters, A. (2014). The competition between legal orders. *International Law Research*, 45–65.

Peters, A. (2016). Global animal law: What it is and why we need it. *Transnational Environmental Law, 5*, 9–23.

Ritvo, H. (2007). On the animal turn. *Daedalus, 136*, 118–122.

Sueur, C. (2016). La fuite de la recherche biomédicale sur les primates en Chine: quelles implications éthiques? *Droit animal éthique & sciences: Revue trimestrielle de la Fondation LFDA, 90*, 19.

Willett, W., Rockström, J., Loken, B., Springmann, M., Lang, T., Vermeulen, S., et al. (2019). Food in the Anthropocene: The EAT–Lancet Commission on healthy diets from sustainable food systems. *The Lancet, 393*, 447–492, Available at: https://doi.org/10.1016/S0140-6736(18)31788-4.

Anne Peters is Director at the Max Planck Institute for Comparative Public Law and International Law in Heidelberg, Professor at the Universities of Heidelberg, Freie Universität Berlin and Basel, and a William W. Cook Global Law Professor at the University of Michigan. She has been a member of the European Commission for Democracy through Law (Venice Commission) in respect of Germany (2011–2015) and served as the president of the European Society of International Law (2010–2012) and of the German Association of International Law (DGIR, since 2019). Her current research interests relate to public international law including its history, global animal law, global governance and global constitutionalism, and the status of humans in international law.

Part I
Historical Foundations

Chapter 2
Rights of and Over Animals in the *ius naturae et gentium* (Sixteenth and Seventeenth Centuries)

Annabel Brett

Abstract The chapter examines different theological and philosophical paradigms of rights in the early modern period. It shows that, contrary to initial appearances, animals were not totally excluded from any kind of right, and that violence against them was not always regarded as legitimate. Remarkably, one of the founders of the discipline of the law of nature and law of nations (*ius naturae et gentium*), Samuel Pufendorf (1632–1694), acknowledged *animal pain*—although he did not translate this acknowledgment into a moral wrong of doing violence to animals, or grant animals moral rights.

De jure naturae et gentium, 'The law of nature and of nations', is the title of Samuel Pufendorf's eight-volume masterpiece of philosophical jurisprudence, first published in 1672. It provides the tag by which an entire discourse is known, one that dominated legal philosophy at European universities for over two hundred years. Pufendorf's Protestant articulation of its principles was pivotal both for transmitting it to the Eighteenth Century and for giving it a history, which in his eyes began with his fellow-Protestant Hugo Grotius. In fact, however, its roots stretch back to the early Sixteenth Century, to the lawyers whom Philip Melanchthon gathered around him at Wittenberg and (more importantly for the future structure of the discourse) to the Catholic scholastic theologians who were originally based at Salamanca in Spain but subsequently spread out over the whole of Counter-Reformation Europe. In the confessional conflict that would burn throughout the

---

Revised version of the original published article "Rights of and over Animals in the Ius Naturae et Gentium (Sixteenth and Seventeenth Centuries)" by Annabel Brett, American Journal of International Law Unbound, Volume 111, 2017, pp. 257–261. The original article was published as an Open Access article, distributed under the terms of the Creative Commons Attribution licence (http://creativecommons.org/licenses/by/4.0/).

A. Brett (✉)
University of Cambridge, Faculty of History, Cambridge, UK
e-mail: asb21@cam.ac.uk

A. Peters (ed.), *Studies in Global Animal Law*, Beiträge zum ausländischen öffentlichen Recht und Völkerrecht 290,
https://doi.org/10.1007/978-3-662-60756-5_2

Sixteenth and Seventeenth centuries, theologians on all sides used law to define the space of the political, and used the idea of natural law to underpin that space, even while shaping it differently according to their divergent narratives of sin and redemption. While the *ius naturae et gentium* was an academic genre, therefore, its content was not. It was a theory and a legitimation of the state, and the arc of its reasoning from nature to the nations ran through the institution of political power. The state and its power are constructed not so much upon right (*ius*) per se, but on the potential for the violation of right (*iniuria*), and the demand for such violation to be vindicated, by law or ultimately by war. At its very barest—although this is to traduce the richness and complexity of the discourse—the *ius naturae et gentium* is a theory of legitimate violence. When it comes to animals, what we find is that they are systematically excluded from the potential to suffer violation of right and therefore from political space and political justice. As we shall see, however, this did not always mean that they were totally excluded from any kind of right or that every act violence against them was always legitimate.

The *ius naturae et gentium* was premised upon an idea of human nature. In the period with which we are concerned, very few doubted—and if they did, it was often (although not always) for the sake of deliberate paradox—that there was such a thing as distinctively human nature, and that this distinction consisted in the natural possession of the capacity to reason. Such conviction was the combined heritage of Christian theology and of (most) classical philosophy, as well as scattered pronouncements in the civil and canon laws. The important point for our purposes is that this heritage prompted scholars in all these three inter-connected disciplines to think through human nature in relation to animal nature. Thus, Scripture in the Book of Genesis directly connected man's nature as the image of God with the *dominium* that God gave man over animals.[1] Aristotelian philosophy (the dominant philosophy of the universities) conceived the faculty of reason as something that human beings possess on top of a whole range of capacities and associated actions that humans have in common with animals. In civil law, the Roman jurist Ulpian likewise posited a natural *ius* that is common (*commune*) to both humans and other animals; he mentioned as belonging to it the union of male and female, which human beings call marriage, and the procreation and rearing of children.[2] For their different reasons, theologians, philosophers and lawyers were almost universally reluctant to deny that the natural capacities and behaviours of animals had any normative force whatsoever. The question was how to square that reluctance with their shared insistence that no animal could make any political claim on any human being.

The Catholic scholastics built their understanding of natural law upon the account offered by the Thirteenth-Century theologian Thomas Aquinas in his *Summa theologiae*. For Aquinas, all law is the work of reason, and natural law is the participation of human reason in God's eternal law. The characteristic of reason is the ability to conceive a good, which is accompanied by the ability to *choose* a good

---

[1]Genesis I. 26.

[2]Mommsen/Krueger/Watson, *The Digest of Justinian* 1985, Vol. 1, Book I, 1.1.3.

and to act purposively to attain that good. The capacity for such action—free-will, in short—is characterised as *dominium* over one's own actions, and only human beings among all terrestrial creatures have it. It is this self-*dominium* that grounds the original *dominium* that God gave man over the other creatures in Genesis I. 26. According to Aquinas, this original *dominium* was granted for the sake of use, by which he mean the purposive ordering of a thing towards a chosen good, for example of food to stay alive. Later scholastics accepted his argument that animals are incapable of use just as they are incapable of *dominium*, because both demand purpose and purpose depends upon rationality. Given that, increasingly, scholastic philosophers of the period conceived a broad equivalence between *dominium* and right (*ius*), the logical consequence of this line of reasoning was to push animals out of the sphere of rights *entirely*: to make them purely used rather than users, incapable of suffering any violation of right and therefore entirely outside any relations of justice. There would therefore appear to be no limit upon the violence that could legitimately be done to them. An extreme example was the Jesuit Luis de Molina (1535–1600) remarking that animals are incapable of suffering *iniuria* and that therefore no more wrong is done to an animal in killing it than in snapping a twig off a tree.[3]

However, other scholastics were uneasy with the proposition that natural animal lives, equally created by God, had no normative value at all. The agreed exclusion of animals from the phenomenon of *dominium* did not mean, for all scholastics, that animals had no rights at all. Some were prepared to keep *ius* separate from *dominium*, at least to some extent, and thereby to see animals as possessing rights in a different sense: not rights *over* things, but rights *to* the activities and goods required for the flourishing of their particular nature. Again this stemmed ultimately from Aquinas, who excluded animals from rational participation in eternal law but allowed them an instinctual participation insofar as they instinctively pursue those things that are good for them. Moreover, a range of the goods that humans are commanded to pursue by natural law are also shared with animals, for example the basic good of self-preservation. The Dominican Domingo de Soto (1485–1650) accordingly allowed 'all things' a natural right to pursue their self-preservation.[4] This might have no implications at all for the rightfulness of human treatment of animals: it might simply mean that animals were behaving in a naturally rightful manner in running away, for example, but not that any wrong was done to them in capturing them. The Jesuit Juan de Salas (1553–1612), however, argued for such rights within an Aristotelian teleology wherein plants are made for animals and animals made for man:

> one should concede to animals – yes, and even to inanimates – right in the sense of what is rightful, or a kind of faculty of doing something, the use of which it would be an injustice to interfere with. For they demand, as if by their own proper right, the things that are naturally

---

[3]de Molina, *De iustitia et iure* 1614. See also Brett, 'Is there any place for environmental thinking in early modern European political thought?' 2018, 23-42.
[4]de Soto, *De iustitia et iure* 1967-1968, Lib. IV, q. 2, a. 2; Lib IV, q. 7, a. 1.

due and proportioned to them, so that they may exist in a good state, and be preserved, and serve the uses of men for whose conveniences they were brought forth.[5]

In this conception, then, justice and rights can exist within a user-used relationship; indeed, it is precisely this God-given relationship, at least in part, which argues for the right. There is thus some limit on human behaviour towards animals, despite the fact that Salas still refused animals any *dominium* and any capacity to suffer *iniuria* in the strict sense.

If we turn now to Protestant natural jurisprudence, we find the topic of animal rights inflected by the different way in which they conceive of natural law. Instead of the focus on individual agency that characterises the Catholic understanding, natural law in the Protestant tradition centrally regulates human relations with others. The *ius naturae* is accordingly the law of natural society. Joining hands with humanist civil jurisprudence in this respect, the Protestant tradition sees *ius* as an inter-personal phenomenon, a relational quality that only exists where there is society or community. It demands the kind of other-regarding behaviour (what would later be termed sociability) that only human beings, as rational creatures, are capable of. Thus the society argument would seem to exclude animals just as surely as the *dominium* argument, and the majority of humanist and Protestant jurists regarded any idea of a *ius naturale* common to human beings and animals as, quite simply, a mistake.[6] Some, indeed, were more sympathetic to the idea that animal agency could be lawful, or that we could talk of right in respect of them. Thus the French humanist lawyer Jacques Cujas (1522–1590) held that animals have been taught *ius naturale*, and follow it equally as do humans, in such things as rearing their young. But all the same, he explicitly ruled out any 'community of right' between animals and humans: there is no shared juridical space, and thus no space for justice or injustice, between them.[7]

In a deliberately controversial early essay on *ius naturale, gentium et civile* published in 1584, Alberico Gentili, Regius Professor of Civil Law at Oxford, countered the humanist argument from community by suggesting that *natural right*—as opposed to right under the law of nations, or under civil law—is possessed by both human beings and animals independently of any community, in the natural activities of their own daily lives: walking, sleeping, running, eating.[8] He recognised, however, that the key question was not simply possession of right, but the potential for its violation (*iniuria*). To the argument that animals do not have rights because they are not capable of violation of right, Gentili responded that there are two kinds of *iniuria*. The first, and central, sense, is that of a deliberate violation, which requires a mind (*animus*); Gentili characterised it as a kind of contumely or contempt. But in a second sense, we can say that everything that happens without

---

[5]de Salas, *Tractatus de legibus* 1611, tract. 13, disp. 2, sect. 2, fo. 35.

[6]The best discussion of the debate is in Scattola, *Naturrecht* 1999, 161-78.

[7]Scattola, *Naturrecht* 1999, 168-69.

[8]Gentili, *Epistolarum ac lectionum libri* IV 1583-4, Lib. III, cap. 1 *De jure naturali, gentium, & civili*, 342-43 [recte 144-5].

right, *sine iure*, also happens by *iniuria* as a kind of unright or nonright, and animals as well as humans are capable of this. The first can only be committed by humans. By contrast, animals, which lack *animus*, can only violate each other's rights (and presumably the rights of human beings too) in the second way. Gentili did not directly say, in this work, whether humans can contumeliously violate animal rights. However, in his later work *On the Law of War* (1598), Gentili addressed the treatment of captives in war and directly paralleled the treatment of slaves and animals, both equally without any rights against their captors either under civil law or under the law of nations. The implication is that they are both confined to the sphere of natural right in the same way, but that nevertheless both have some juridical claim against their captors. His initial point was one about kindness rather than justice, and his example is the Athenians who allowed their animals rest, pasture and even burial after 'the long labours of life'. The argument takes a more legal turn, however, when he invoked 'the law of God', i.e. Deuteronomy: 'Thou shalt not muzzle a threshing ox'.[9] If this is not strict *iniuria*, it is nevertheless some kind of affront to justice in a broader sense, a sense that includes both human beings and animals.

Two towering figures of the Protestant Seventeenth Century followed Gentili in positing animal rights, but without the consequences for human treatment of them that Gentili drew. In his early work *De iure praedae*, Hugo Grotius (1583–1645) recognised a right (or at least the rightfulness) of self-preservation in all animals, human or otherwise. However, he placed this right of pursuing one's own good explicitly prior to any recognition of the other's good, and thus prior to any justice properly so-called and to the natural society that depends upon it. For Grotius, this requires reason, which animals do not possess. He did accord them a glimmering of regard for others, for example in the way that they care for their young, but held that this faint sense is not enough for justice or for natural society.[10] Animal rights are therefore outside the sphere of justice; they place no constraints on the rights of human beings. Equally, Thomas Hobbes, at least in *De cive* (1642), placed animals and humans in the same natural juridical space when he said that a man will kill an animal with the same right that an animal kills a man.[11] But this space is the 'condition of nature', which Hobbes famously equated with a condition of war, in which again there is no justice, or at least no justice in effect.[12] No one—no human being, no animal—can be convicted of violating another's right in the condition of nature. Human beings can escape this terrible situation by covenanting to create a civil state, but animals remain forever in the condition of nature. They can never have any right not to be killed or otherwise used for any purpose, albeit neither can they ever be obliged not to kill human beings.

---

[9]Gentili, *De jure belli libri tres* 1933, Vol. II: Translation, Bk III, Ch. 9.

[10]Grotius, *Commentary on the law of prize and booty* 2006, 21-28.

[11]Hobbes, *De cive* 1998.

[12]Hobbes, *Leviathan* 1996, Ch. 13.

Despite their differences, Gentili, Grotius and Hobbes were all in these works indebted to a distinct strand within late renaissance philosophy which deliberately denied, or even inverted, the traditional theory of the natural superiority of human beings over animals based on the supposedly natural and exclusive human possession of reason.[13] In its neo-Epicurean mode, shared by the early Grotius and Hobbes, this served to place human beings and animals equally in a naturally lawless condition. But in another version—drawing on other strands of Hellenistic philosophy, such as skepticism, ancient vegetarianism, and the Stoic repugnance towards anger and cruelty and other destructive passions—the emphasis was rather on the positive qualities that animals shared with human beings. Gentili referred to some of this material to confirm his thesis that animals were included in the *ius naturale*.[14] Ideas of the reason and virtue of animals were powerfully articulated by writers like Michel de Montaigne, especially in the *Apology for Raymond Sebond*, and Montaigne's follower Pierre Charron, whose works enjoyed huge success all over Europe.[15] Although the strategy of inversion was primarily deployed as an invitation to moral reflection and inner freedom, this style of philosophy was also marked by its insistence on the concrete practice of our lives and our behavior towards other beings. It left a deep impression upon Samuel Pufendorf, who engaged both approvingly and disapprovingly with Charron's *De la sagesse* throughout the first books of *De jure naturae et gentium*.

Pufendorf's legal philosophy rested upon a distinction, inherited ultimately from Jesuit scholastics such as Francisco Suárez, between 'natural entities' and 'moral entities'. While the former are the result of natural processes, the latter are 'imposed' by the free will of a rational agent: God, in the first instance, and human beings thereafter. The created 'moral' world, which includes persons, rights, obligations, statuses, powers, and all values, excludes animals, since for Pufendorf its condition, rationality, was an exclusively human characteristic, 'whatever Charron... has maintained to the contrary'.[16] He stressed that moral entities distinguish all that is decorous and civilised in the life of human beings from the brute life of animals, underpinning the point with a complex psychological account of the working of such moral entities upon human passions and actions. The direct corollary of this position, however, was that the superiority of human life was not a function of reason alone. This was his debt to Charron and to Hobbes. Without moral entities, reason was a mere cunning; morality is 'imposed', not natural.[17] But while Pufendorf endorsed some of the positive evaluation of animal lives as lacking the vices that characterize

---

[13]See the contribution of Anna Becker, chapter 3, in this volume.

[14]Gentili, *Epistolarum ac lectionum libri* IV 1583-4, Lib. III, cap. 1 *De jure naturali, gentium, & civili*, 346-47.

[15]de Montaigne, *An apology for Raymond Sebond* 1987; Charron, *De la sagesse* 1986. Two translations of the latter work into English were made at the beginning and end of the seventeenth century, by Samson Lennard and George Stanhope.

[16]Pufendorf, *De jure naturae et gentium libri octo* 1934, Vol. II: Translation, Book I, Ch. 3, n. 1. The reference is to Book I, Ch. 34 of *De la sagesse*.

[17]Pufendorf, *De jure naturae et gentium* 1934, Book II, Ch. 1-2; Book I. Ch. 6.

human beings, the same theory of moral entities meant that such lives had no moral quality, nor did animals have any moral right to live them. God did not impose any natural law on both humans and animals; what they had in common was certain regular natural behaviours, like rearing young, but the fact that God had created these regularities did not make them law, and any natural goodness inherent in them was not sufficient for moral status.[18]

When it came to natural human rights over animals, Pufendorf, like Montaigne, rejected as a piece of human vanity the view that the only reason for other creatures' existence was to serve human beings. Nevertheless, he argued that God had created man with needs which could not be met without using animals, a use which therefore must have been morally licensed by God and which by the same token could not involve the violation of right (*iniuria*).[19] But he then paused over the question whether that use legitimately extended to slaughtering them. After all, humans could use animals for food in other ways, for example by milking them; they did not *have* to eat them. The crucial point is that, unlike plants, animals suffer agony (*cruciatus*) in being killed. Pufendorf here rehearsed at striking length all the late renaissance philosophical arguments against doing violence to animals. His response was curiously indirect. He approved temperance and frugality in eating. But from the assertion that God had imposed no law in common between human beings and animals, and thus no command to cultivate mutual society, Pufendorf drew the conclusion that there was a state of war between them, and thus that neither side could commit *iniuria* on the other. Therefore, human beings do not commit *iniuria* on animals when they kill them, either for food or for any other reason such as keeping the population down. He ended with a caution, however, against abuse of animals as an abuse of God's creation.

In his popular abridgement of *De jure naturae et gentium*, entitled *De officio hominis et civis*, Pufendorf kept only the first of the above arguments. Remarkably, however, he also kept the mention of animal pain: '(. . .) harmless animals which it is not wrong for men to kill and consume, even though they die in pain (*dolore*).'[20] As we have seen, Pufendorf's whole system depends on separating out a natural evil, like pain, from moral wrong. And yet his mention of it ruptures the smooth flow of legitimation. In allowing animal pain to break the surface of his text, Pufendorf preserved some of the late renaissance attack on human complacency even while he denied that it translated into animal rights against humans. In terms of the preceding *ius naturae et gentium*, this acknowledgement of *pain* is (to the best of my knowledge) unique.

---

[18]Pufendorf, *De jure naturae et gentium* 1934, Book II, Ch. 3; Book I, Ch. 2.

[19]Pufendorf, *De jure naturae et gentium* 1934, Book IV, Ch. 3.

[20]Pufendorf, *On the duty of man and citizen* 1991, Book I, Ch. 12.

References

Brett, A. (2018). Is there any place for environmental thinking in early modern European political thought? In K. Forrester & S. Smith (Eds.), *Nature, action and the future. Political thought and the environment* (pp. 23–42). Cambridge: CUP.

Charron, P. (1986). *De la sagesse*. Paris: Fayard.

de Molina, L. *De iustitia et iure* Tract. II, disp. 3, n. 6 (ed. Mainz 1614).

de Montaigne, M. (1987). *An apology for Raymond Sebond*, (Michael Andrew Screech, Trans.). (Eds.), Harmondsworth: Penguin.

de Salas, J. (1611) *Tractatus de legibus*. Barcelona.

de Soto, D., *De iustitia et iure* (facs. ed. Salamanca 1556 with parallel Spanish translation, Madrid: Sección Teólogos Juristas, Instituto de Estudios Políticos 1967–1968).

Gentili, A., *Epistolarum ac lectionum libri* IV (London 1583–1584).

Gentili, A. (1933). *De jure belli libri tres* (Vol. 2). Oxford: Clarendon Press.

Grotius, H. (2006). *Commentary on the law of prize and booty* (Gwladys L. Williams, Trans.). Martine Julia van Ittersum (Eds.), Indianapolis: Liberty Fund

Hobbes, T. (1998). *De cive* (Richard Tuck and Michael Silverthorne, Trans.). (Eds.), Cambridge: CUP.

Hobbes, T. (1996). In R. Tuck (Ed.), *Leviathan*. Cambridge, CUP.

Mommsen, T., Krueger, P., & Watson, A. (Eds.). (1985). *The Digest of Justinian* (Vol. 4). Philadelphia PA: University of Pennsylvania Press.

Pufendorf, S. (1991). *On the duty of man and citizen* (James Tully, Trans.). Michael Silverthorne (Ed.), Cambridge: Cambridge University Press.

Pufendorf, S. (1934). *De jure naturae et gentium libri octo* (Vol. 2). Oxford: Clarendon Press.

Scattola, M. (1999). *Das Naturrecht vor dem Naturrecht. Zur Geschichte des "ius naturae" im 16. Jahrhundert*. Tübingen: Niemeyer.

Annabel Brett is Professor of Political Thought and History at the University of Cambridge. She is the author of *Liberty, right and nature. Individual rights in later scholastic thought* (1997) and *Changes of state. Nature and the limits of the city in early modern natural law* (2011), as well as of a number of articles on the history of natural law, human rights and international law. A major recent theme of her research is the way in which geographical space and non-human nature have historically been conceived in relation to human beings and political formations, and how those issues are theorised in contemporary political thought.

Chapter 3
On Women and Beasts: Human-Animal Relationships in Sixteenth-Century Thought

Anna Becker

Abstract The chapter traces how early modern writers of political theory, often in their comments on Aristotle, viewed the relationships between some animals and humans, notably in the household. Remarkably, not all authors drew a sharp contrast between the human male on the one side, and disenfranchised women, slaves, and animals, on the other. Some writers did not view humans as completely alienated from their animal nature. Thus, early modern writers' contemplation of the human animal and the fluidity between nature and culture might inspire current reflection on animal welfare and rights.

We are used to the view that historically 'what counted as fully human always depended (. . .) on a sharp contrast with "the animal"'. As a consequence, we might think that 'women and slaves, in being denied full humanity, were therefore necessarily partaking in animal nature'.[1] This chapter questions the view that early modern philosophers worked with a simple analogy of women and beasts. The chapter traces how some early modern thinkers defined the relationship of human beings to animals generally, and, more particularly, how they saw the relationship of women, slaves, and animals in the human household. The picture presented, while being far from complete, shows that fifteenth- and sixteenth-century thinkers had

Revised version of the original published article "On Women and Beasts: Human-Animal Relationships in Sixteenth-Century Thought" by Anna Becker, American Journal of International Law Unbound, Volume 111, 2017, pp. 262–266. The original article was published as an Open Access article, distributed under the terms of the Creative Commons Attribution licence (http://creativecommons.org/licenses/by/4.0/)

[1]Tanasescu, *Environment, political representation, and the challenge of rights: speaking for nature* 2016, 65.

A. Becker (✉)
Aarhus University, Department of Philosophy and History of Ideas, Aarhus, Denmark
e-mail: anna.becker@cas.au.dk

© The Author(s) 2020
A. Peters (ed.), *Studies in Global Animal Law*, Beiträge zum ausländischen öffentlichen Recht und Völkerrecht 290,
https://doi.org/10.1007/978-3-662-60756-5_3

nuanced arguments to offer, when discussing the relationship of human animals to non-human animals, as well as a complex reading available on what we would regard as the relationship of 'nature' to 'culture' more broadly speaking.[2] What is more, as we shall see, the equation of women with animals and slaves was not something that we commonly find in sixteenth-century philosophical treatises, which might lead us to rethink our own ideas on the relationship of one disenfranchised group with the other.

In the first book of the *Politics,* Aristotle described how we come to be political animals 'in a sense in which the bee is not or any gregarious animals'.[3] Human beings differed from animals because they possessed speech and reason.[4] Those qualities made human beings form cities and households, both very 'human' forms of society. Even gregarious animals were not thought to come together in a well-ordered family life. Such a well-ordered human household, Aristotle argued, was formed out of two original societies or associations of command and obedience. It consisted of the society of husband and wife for the procreation of future generations, and of the society of master and slave for the day-to-day securing of subsistence. When Aristotle introduced his thoughts on households he underlined his argument with a quote from Hesiod's *Work and Days*:

From these two partnerships then is first composed the household, and Hesiod was right when he wrote 'Get first a house and a wife and an ox to draw the plough'. (The ox is a poor man's slave). This (...) is the household, the members of which Charondas calls 'bread-fellows' and Epimenides the Cretan 'stable-companions.[5]

What was human was characterised by distinguishing it from the animal, but the animal was seen as part of the very human household. In early modern Europe, philosophers took the Aristotelian argument seriously and discussed its implications. In his 1587 commentary on the *Politics,* the Ferrarese professor for natural philosophy, Antonio Montecatini (1537–1599), took up Hesiod's quote. He wrote that a 'perfect' household consisted of two societies, the 'marriage society and the master-servant society'. For a master-servant association to be perfect, then, it was 'sufficient' (*sanum*) as Hesiod had written, to have 'at once with the wife an ox; and certainly Hesiod meant the ox to be the servant. Because the ox plays the role of the servant in the households of the poor and especially that of the peasants. By all means it is the farmer's partner and aid (*socius et minister*).'[6]

---

[2]In recent years, animal-human studies have emerged as an influential field of research in the humanities and in the social sciences. For an overview of the links between feminist scholarship and human-animal studies, see Birke, 'Intimate Familiarities?' 2002, 429–436. For one example of a nuanced depiction of animal-human relationship in early modern European texts, see Fudge, *Brutal Reasoning: Animals, Rationality, and Humanity in Early Modern England* 2006.

[3]Aristotle, *The Politics* 1992, 1253a7, 60.

[4]Ibid.

[5]Ibid., 1252b9, 58.

[6]Montecatini, *In politica hoc est in civiles libros Aristotelis Antonii Montecatini ferrariensis progymnasmata* 1587, 29.

The very human institution of the household, for Montecatini, could be called perfectly instituted when it consisted of a fellowship of human beings with animals. Animals hence were able to take the place of a human being. In this case the animal was characterised as *socius*, as partner, and as associate in the family realm, which it perfected. Montecatini made certain that his readers understood that, in its function as servant, the ox did not only share in the 'abstract' conceptual space of the household but also in its concrete and material space. Epimenides had called the household members 'stable companions', Montecatini argued, because 'they were eating at the same crib and from the same table'.[7] The perfect household was a space in which animals worked in humans' stead, a space in which human beings shared with animals their food and their lives.

It is of course true that Aristotle had talked about the slave-master relationship as a household relationship that was characterised by a sharp hierarchy. The slave was an unfree human being without any political rights, dependent on the will of the master (who needed however treat the slave in concordance with the overall good of the household).[8] This indeed might suggest that substituting the animal for the slave does not make a difference in terms of their juridical and philosophical positions. This is however not to the point. A slave, in the Aristotelian universe, is still a human being and seen from that perspective, exactly not an animal. We should also further note that in the context Montecatini was writing it is clear that he thought of the agricultural family of his own time, in which servants, not slaves were the human help. That Montecatini called the animal *socius et minister*, finally, shows that what was focused on here was not a 'degradation' of a human being into the position of an animal, but an 'elevation' of the animal into the position of a human being, as a companion to a human in the pursuit of every-day life.

While the animal could stand in for a slave or servant, early modern Aristotelian commentators were careful not to conflate 'woman' with 'animal', or, more precisely in this case, wife with ox. According to Aristotle in the *Politics*, it was a sign of barbarism to treat wives as slaves. Early modern commentators extended this further and showed that even worse was when wives were used as animals. Montecatini wrote:

> In our time chiefly the wives of the Germans and the Helvetians serve their husbands, while they travel with them, and they carry heavy loads. It is as if they were slaves, or rather, as if they were mules. They carry inhuman burdens. The laws of the Mohammedans and their worshippers have wives not for slaves but for animals.[9]

Protestants and Muslims hence showed their considerable difference to those of the right faith in the way they treated their wives—as animals. Equating wives with animals was a sign of a life not lead morally well. It showed an unfavourable difference in religion, in civilisation, and in culture. It was a marker for strangeness. Even the most foundational unit of human life, the household, for Aristotelians was

---

[7]Ibid., 31.

[8]See Aristotle, *The Politics* 1992, 1254a9, 65.

[9]Montecatini, *In politica* 1587, 27.

not ordered according to one universal nature. 'Culture' determined what form family life took, cultural identity formed the relationships at the level of individual households.

This idea about civilisation, culture, and the 'right' life could also be turned against the very culture and religion that it ordinarily defended. One example of this can be found in 'El Inca' Garcilaso de la Vega's *Comentarios Reales de los Incas* (1606). 'El Inca', born 1539 in Peru as the son of an Inca princess and a Spanish conquistador, wrote the most famous history of the conquest. His work was not only a historical account but also a political treatise that defended the authority of the Inca and argued that mestizos were far more capable to rule the 'New World' than Spaniards could ever be.[10] Throughout his work Garcilaso played on classical references, using tropes from Greek philosophy and Roman rhetoric, and subverted them to great effect. He argued that the Inca had erected a second Rome (*otra Roma*), and had managed to do so without the exploitation of animals. In vivid detail he described the instance oxen first came to Peru in 1551. The Inca observers, he wrote, 'said that the Spaniards, who were drones, and would not work themselves, had made these great Animals labour and doe that work which they ought to have performed themselves'.[11] Garcilaso here turned up-side down the known trope of animals as servants. Using an animal to work on the field did not signify the cultural superiority of the conquistadors but in the eyes of the Inca rather the opposite: corruption, idleness, and laziness, and as such attributes normally assigned to the Inca by the Spanish.

Returning to the topic of the animal-woman relationship, my point here is to show that early modern humanists did not operate with a simple binary in which 'disenfranchised' beings, i.e. animals, women, and slaves, were on one side and the human male (as creation's crown) on the other. Early modern Aristotelian philosophy clearly saw both women and slaves as human beings, and as such different to animals. Neo-stoic thought had emphasised that all human beings were bound together in a cosmopolitan fellowship. From this shared *humanitas* however, did not flow civil rights; and the idea of the common humanity was not at all incompatible with strict hierarchical thinking in legal terms. In early modern European cities it was *status* that determined the civic rights of every person. The laws, privileges, and duties of early modern personhood were dependent on a myriad of different categories. This was not a matter of 'man' or 'woman'. Rather, *patres familias*, mothers, widows, married men, unmarried women, servants, and maids all had different civic standings.[12] Granted, all human beings were different from animals, but the extent of the difference, or the distance that separated one specific

---

[10]Becker, 'Fragile Männlichkeiten?' 2017, 249-261.

[11]de la Vega, *The royal commentaries of Peru, in two parts* 1688, 9, XVII, 378.

[12]On issues of legal standing, status, and citizenship of women in early modern Europe see e.g. Kirshner, *Marriage, Dowry, and Citizenship in Late Medieval and Renaissance Italy* 2015 and Kuehn, *Family and Gender in Renaissance Italy, 1300-1600* 2017.

animal from one specific human being shifted, both according to the legal position of the human being, and according to the hierarchical position of the animal.

Indeed, the Latin word *animal* was used more frequently to describe human beings than non-humans. Sixtus Birck (1501–1554) defined human beings as 'social animals, bipeds', clearly here underlining what human and non-human animals had in common rather than emphasising what divided them.[13] When early modern humanists wanted to make a particular point about non-human animals (rather than speaking of 'creatures' in a broad sense), they used the term *bellua,* beasts, that were then divided into tame (domesticated) and wild beasts, which could be divided into even more subcategories.[14] The Holy Script, too, divided animals into 'the fish of the sea', 'the foule of the heaven', and 'the beast of the fielde'.[15] Even between animals, there was thus a hierarchy in terms of their power relationship to human beings. Keeping in mind that human society was deeply hierarchical, it is fitting that 'animals' also were split into many different sub-categories.

It is in the context of the wild against the tame that we find, as the sixteenth century drew to a close, the equation of women with animals. In the *Francogallia* (1573), a work that argued that France traditionally had a constitutional past and that French citizens had the right to overthrow any king who turned out to be a tyrant, the author, François Hotman, described female rulers as *indomitus*, that is, as 'untamed' animals as well as 'unbridled beasts'.[16] Hotman (1524–1590) thereby indicated that women who aimed to rule behaved like wild beasts, irrational, dangerous, and blood-thirsty. A male tyrant was bad enough; a female tyrant was nothing less than a raging animal. She had left her natural and well-ordered domain, the household, for a brutish habitat. This trope, however, was far from persistent. In the Protestant imagery of the eighteenth century it often was the woman who was seen as the civilising influence over men who, in turn, were described as having sexual appetites like wild beasts.[17]

Renaissance authors discussed in more detail what 'made' wild animals become tame. Some ancient authors had actually suggested that the current state of animal-human relations was not simply due to the natural order. They argued that in the past there must have been a sort of pact between animals and human beings, which made animals obey human rule. In Lucretius' *De rerum naturae*, widely received in Renaissance political thought and supporting a non-anthropocentric world view, readers could find an example on how that pact might have come into being. Lucretius had argued that the dangers emanating from sharing life in the wild with beasts must have been the greatest motivator for human beings to originally form associations and so escaping a 'brutish', uncivilised situation. Hand in hand with the

---

[13]Birck/of Rotterdam/Amerbach/Maturanzio, *De officiis Commentarii,* 1562, 9v.

[14]See e.g. Montecatini, *In politica* 1587, p. 172.

[15]Genesis 1:26, 2:19. See also Shannon, 'Poor, Bare, Forked' 2009, 173-74.

[16]Hotman, *Francogallia* 1972, 484.

[17]See e.g. Hull, *Sexuality, State, and Civil Society in Germany* 1996. A close investigation of the animalisation of the sexual appetites from a gendered perspective still remains to be written.

development of civilisation came the pact between animals and humans, an act of reciprocity in which animals exchanged what was useful for tutelage.[18] In his 1570 commentary on *De rerum naturae*, Denys Lambin (1520–1572) informed us that in exchange for protection, human beings profited from 'oxen, goats, horses, and asses' as beasts of burden. The animals provided their physical strength and served as means of transportation, as well as giving human beings their furs as vestments.[19]

With his description of a ruler-ruled relationship between animals based on reciprocity, Lambin actually came close to contemporary descriptions of monarchical rule, particularly that of an absolute ruler. The most important apologist for absolute rule, Jean Bodin (1530–1596), had, in his *Six livres de la republique* (1576), argued that this was exactly what a monarch ought to do: provide care, protection, and tutelage in exchange for the absolute obedience of the subjects.[20]

Following Lucretius amongst others, some Renaissance thinkers argued strongly that human beings could not without problems be called 'superior to beasts'. Renaissance writers, contrary to how their thought is often understood, were not at all only concerned with placing the individual in the centre of the universe. On the contrary, praise of the exalted position that human beings held in the universe was often mixed with a reminder for the readers that with the Fall human beings had brought on themselves absolute misery. A famous example of a writer denying human superiority over animals was Michel de Montaigne (1533–1592). In the longest chapter of the *Essais*, the *Apologie de Raimond Sebond* (1580), Montaigne did his best to show that human beings had no reason to claim that they were inherently better than animals.

> Beasts are born, reproduce, feed, move, live and die in ways so closely related to our own that, if we seek ... to raise our own status above theirs, that cannot arise from any reasoned argument on our part. Doctors recommend us to live and behave as animals do.[21]

Montaigne also argued that animals had prudence, even a sense of justice, in the classical Ciceronian phrasing of 'rendering everyone his due'.[22] Animals were intelligent, they were able to learn even complex things, they communicated with each other, they knew how to administer medicine, and they certainly were better at counting than human infants. Montaigne even denied the singularity of what was so often understood as the most human aspect of the human condition, namely the political community. What 'form of body politic [has been] more ordered [...] than that of the bees?' he asked. He went even further and claimed that

> man must be restrained with his own rank within the boundary walls of this polity; the wretch has no stomach for effectively clambering over them: he is trussed up and bound subject to the same restraints as the other creatures of his natural order.[23]

---

[18]Lucretius, *On the nature of things* 1948, V, II 860-77, 214-15.

[19]Lambin, *T. Lucretii Cari De Rerum Naturae Commentarii* 1570, 479.

[20]Bodin, *Les Six Livres de la République* 1986, I, 6, 141.

[21]de Montaigne, *The Complete Essays* 2003, 524.

[22]Ibid., 525.

[23]Ibid., 514.

He thus showed that the natural habitat of human beings was no different in character from the natural habitat of animals. Thereby he situated human beings very directly into their own nature: The polity was to humans what the jungle or steppe was to animals: their natural habitat. Animals were not easily able to leave their natural surroundings without giving up their nature, but neither were human beings. Beasts and human beings might live in different places, but they were still obeying the same rules in relation to these places. This shared condition did not allow for the claim of superiority of the one over the other.

Some 50 years earlier Niccolò Machiavelli (1469–1527) had actually suggested a way that human beings could 'climb over' the wall of the polity. He praised Achilles and 'many other ancient rulers' who were raised by Chiron the centaur, because 'having a mentor who was half-beast and half-man signifies that a ruler needs to use both natures and that one without the other is not effective'. To be successful, Machiavelli argued, a ruler must know 'how to act like a beast' and needed to fashion himself to be like lion and fox.[24] Only if rulers were able to liken themselves to animals they would be able to withstand the dangers governing brought with it and could hope to successfully maintain their governments. Clearly the best ruler needed to know how to be both: human and beast.

The above analysis suggests that early modern thinkers did not have one binary world view, but were thinking of the relationship between some non-human animals and some human beings as changing and shifting. Renaissance thinkers assumed that human beings and animals shared very similar origins. In the *Digest* Ulpian had laid down that the natural law is

> that which nature has taught to all animals, for this law is not peculiar to the human race, but applies to all creatures. Hence arises the union of the male and the female which we call marriage; and hence are derived the procreation and the education of children; for we see that other animals also act as though endowed with knowledge of this law.[25]

As Annabel Brett has shown, the idea that this constituted 'a society of law' with animals was mainly denied by early modern commentators, while, on the other hand, they also never claimed that human beings were completely alienated from their animal nature.[26] While early modern thinkers situated human beings into nature, they also understood that what seemed to be 'nature' might as well be 'culture': Human beings were political by nature, but their polities (or 'states' in modern parlance) were built in a process of civilisation. Different polities had different laws and customs, but this was often thought to reflect the diverse 'nature' of different cities. In this sense, fulfilling one's nature often needed cultivation. This, and the fact that we should rethink the relationship of one disenfranchised group to another, rather than assuming that their concerns might always be alike, might be a way that the distant past can still inspire us to think about pressing issues today.

---

[24]Machiavelli, *The Prince* 1988, XVIII, p. 61.

[25]*Digest*, 1.1.3

[26]See Brett, *Changes of state* 2011, especially this chapter, and her chapter in this volume.

References

Aristotle. (1992). *The Politics* (T. A. Sinclair, Trans.). Rev. T. J. Saunders. London: Penguin.
Becker, A. (2017). Fragile Männlichkeiten? Die Comentarios reales des Inka Garcilaso de la Vega zwischen europäischer Wissenstradition und peruanischer Selbstbehauptung. In S. Richter, M. Roth, & S. Meurer (Eds.), *Konstruktionen Europas in der Frühen Neuzeit* (pp. 249–261). Heidelberg: Heidelberg University Publishing.
Birck, Sixtus/of Rotterdam, Erasmus/Amerbach, Veit/Maturanzio, Francisco/M. T. Ciceronis [. . .] De officiis Commentarii (Basel 1562).
Birke, L. (2002). Intimate familiarities? Feminism and human-animal studies. *Society & Animals, 10*, 429–436.
Bodin, J. (1986). *Les Six Livres de la République.* In C. Frémont, M.-D. Couzinet, & H. Rochais (Eds.), 6 vols. Paris.
Brett, A. (2011). *Changes of state. Nature and the limits of the city in early modern natural law.* Princeton-Woodstock: Princeton University Press.
de la Vega, G. (1688). *The royal commentaries of Peru, in two parts* (Paul Rycaut, Trans.). London.
de Montaigne, M. (2003). *The complete essays* (M.A. Screech, Trans.). London: Penguin.
Fudge, E. (2006). *Brutal reasoning: Animals, rationality, and humanity in early modern England.* Ithaca, NY: Cornell University Press.
Hotman, F. (1972). *Francogallia* (J. H. M. Salmon, Trans.). R. E. Giesey (Ed. and Latin text), Cambridge: CUP.
Hull, I. V. (1996). *Sexuality, state, and civil society in Germany, 1700–1815.* Ithaca, NY: Cornell University Press.
Kirshner, J. (2015). *Marriage, dowry, and citizenship in late medieval and renaissance Italy.* Toronto: University of Toronto Press.
Kuehn, T. (2017). *Family and gender in renaissance Italy, 1300–1600.* Cambridge: CUP.
Lambin, D. (1570). *T. Lucretii Cari De Rerum Naturae Commentarii.* Paris.
Lucretius. (1948). *On the nature of things* (Cyril Bailey, Trans.). Oxford: OUP.
Machiavelli, N. (1988). *The Prince* (Quentin Skinner, Trans.). (Ed.), Russell Price. Cambridge: CUP.
Montecatini, A. (1587). *In politica hoc est in civiles libros Aristotelis Antonii Montecatini ferrariensis progymnasmata.* Ferrara.
Shannon, L. (2009). Poor, bare, forked: Animal sovereignty, human negative exceptionalism, and the natural history of king lear. *Shakespeare Quarterly, 60*, 168–196.
Tanasescu, M. (2016). *Environment, political representation, and the challenge of rights: speaking for nature.* London: Palgrave Macmillan.

Anna Becker is Professor MSO at the Department for Philosophy and History of Ideas at Aarhus University; Denmark. She has published widely on topics in early modern political thought and political culture, including political Aristotelianism, oeconomics and gender, and colonial masculinities. She was awarded the Balzan–Skinner Fellowship for Intellectual History in 2014–2015.

Chapter 4
Animal Colonialism: The Case of Milk

Mathilde Cohen

Abstract The chapter examines 'animal colonialism' and one of its iterations, 'milk colonialism.' Until the end of the nineteenth century—and sometimes well into the twentieth century—the majority of the world population outside Europe neither raised animals for their milk nor consumed animal milk. With the violent colonization of the New World and other territories starting in the sixteenth century, dairying began to spread globally. European settlers did not set out to colonize lands and people alone, they transported with them animals and plants, including lactating animals such as cows and sheep. These living imports not only disrupted local ecosystems, but also relational patterns by altering, sometimes even severing, the breastfeeding relationship between females, be they animal or human, and their young. By propagating and spreading animal milk consumption and depreciating colonized women's practice of breastfeeding, the oppression of humans and animals went hand in hand. This account adds a fascinating dimension to the history of the international law of development.

1 Introduction

Greta Gaard writes that '[t]he pervasive availability of cows' milk today—from grocery stores to gas stations—is a historically unprecedented product of industrialization, urbanization, culture, and economics.'[1] To these factors, I would add

---

Revised version of the original published article "Animal Colonialism: The Case of Milk" by Mathilde Cohen, American Journal of International Law Unbound, Volume 111, 2017, pp. 267–271. The original article was published as an Open Access article, distributed under the terms of the Creative Commons Attribution licence (http://creativecommons.org/licenses/by/4.0/).

---

[1]Gaard, 'Toward a Feminist Postcolonial Milk Studies' 2013, 597.

M. Cohen (✉)
University of Connecticut, School of Law, Hartford, CT, USA
e-mail: mathilde.cohen@uconn.edu

© The Author(s) 2020
A. Peters (ed.), *Studies in Global Animal Law*, Beiträge zum ausländischen öffentlichen Recht und Völkerrecht 290,
https://doi.org/10.1007/978-3-662-60756-5_4

colonialism and international law, the latter understood broadly to include the rules considered binding between states and nations, transnational law, legal transplants, international food aid, and international trade law. Until the end of the nineteenth century, the majority of the world population, especially inhabitants of the American continent, tropical Africa, South East Asia, the Far East, Australia, and the Pacific Islands, neither raised animals for their milk nor consumed animal milk. Humans are unique in the mammalian realm in that they drink the milk of other species, including beyond infancy. With the European conquest of the New World and other territories starting in the sixteenth century, dairying began to spread worldwide—settlers did not set out to colonise lands and people alone; they brought with them their flora, fauna, and other forms of life, including lactating animals such as cows and sheep.

Though limited to a narrow group of animals and products, the story of milk's globalisation may have broader implications for how we understand the genealogy of global animal law. It suggests that animal law may long have been 'global', at least since the modern era, which saw the colonisation of lands in the Americas, Africa, Asia, and Oceania by a few European countries and the accompanying migration of ideas concerning the legal status of animals. There is a fundamental difference between the new and the old global animal law, however. While the contemporary global animal law initiative embraced by this book is an emancipatory movement aiming at promoting animal welfare, the old, colonial animal law was only global for imperialist ends, displaying little concern for the well-being of animals, colonised people, and ecosystems.

Bridging the gap between scholarship on animal colonialism and on imperialism and motherhood, this essay argues that lactating animals became integral parts of colonial and neo-colonial projects as tools of agro-expansionism and human population planning. Due to its disruptive effects on breastfeeding cultures, the global spread of dairying has not only been detrimental for the welfare of animals, but also for humans', especially mothers and their children. I recognise the simplistic aspect of grouping and analysing together disparate epochs, regions, peoples, and animals in an inter-imperial historical vein. I do not mean to imply that these epochs, regions, peoples, and animals belong to a coherent whole, but only that despite their diversity, they have experienced comparable forms of state-building projects centred upon the consumption of animal milk. As an aside, animal protection law and advocacy is often critiqued for its supposed cultural imperialism, but as the following discussion illustrates, it may be that the lack of concern for animal welfare exhibited by legal systems was bequeathed by hegemonic European colonisers.

In what follows, after presenting the notion of animal colonialism, I focus on two of its components, which I call "milk colonialism" and "breastfeeding colonialism" before concluding with a provocative proposal about the international right to breastfeed.

2 Animal Colonialism

Animal colonialism can be defined as a dual phenomenon, consisting, on the one hand, in using animals to colonise lands, native animals, and people and, on the other hand, in imposing foreign legal norms and practices of human-animal relations upon communities and their environments. Beginning with the work of Alfred Crosby on ecological imperialism—in particular his insight that the conquest of the New World was as much a biological one as a political one[2]—studies have accorded domesticated farm animals an instrumental role in the establishment of colonies (or 'Neo-Europes' to use Crosby's words) around the globe. Virginia DeJohn Anderson writes: 'all Europeans (. . .) enlisted livestock as partners in colonization.'[3] By displacing local fauna, altering native weeds, seeds, grasses, and cultivars, ranching and dairying altered New World ecosystems to advance European purposes. This biological invasion disrupted the lives of native peoples, animals, and their environments.

Historically, animals have been integral parts of colonial and imperial projects (I use these terms interchangeably to include more recent instances of neo-colonialism) as essential tools of imperial agro-expansionism. Across time and space, colonists used animals to conquer ecosystems and their inhabitants, from Christopher Columbus who transported horses, cattle, swine, sheep, and goats to Caribbean islands to French settlers who brought cattle to New France starting in 1617, to Dutch settlers who exported their first cows to New York in 1629, to the British who landed with their sheep and bovines on the shores of Australia and New Zealand in the eighteenth and nineteenth centuries. Similar in these disparate endeavours was the idea that the importation of European animals and the destruction of local fauna, flora, and local foodways were justified by the goal of 'improving' agriculture and population health.

Animals and their 'products'—in particular milk, leather, fur, bone, wool, and silk—were and remain constitutive of national identity and imperial power. They operate as tools of domination to control territories, humans, animals, and ecosystems. Animal colonialism also served as a pretext for conquest itself: as the imported cattle multiplied, more grazing land was needed, justifying further expansions. According to colonists, farming established legitimate legal entitlements to the land, which was conceptualised as a *res nullius* (empty thing) remaining common property until put to use. This was the Lockean idea that men acquired civil rights when they appropriated tracts of lands to themselves and used them productively. As Virginia DeJohn Anderson has shown about North America,[4] by making agriculture the sole measure of use, colonists denied native peoples of New England and Virginia such as the Algonquians, Patawomecks, Powhatans, and Wampanoags any claim to the hunting lands essential to their way of life (and of course they did

---

[2]Crosby, *Ecological Imperialism* 1986.
[3]Anderson, *Creatures of Empire* 2004, 97.
[4]Ibid., 6.

farm, but very differently from Europeans—using smaller, unfenced parcels and growing other varieties of crops among other differences.)[5]

Animal colonialism involves not only the migration of animals, but also of the legal status they were accorded in the Old World. Imperial states can recognise or refuse to recognise the legitimacy of multispecies relationships in the regions they conquer, attempting to regulate them or replace them with new ones. In both the civil law and the common law traditions, animals were the personal property or chattel of their human owners and could not possess rights.[6] They were means to human ends. One illustration of how the property status of animals became 'globalised' is the seventeenth century introduction of the crime of animal theft in the Chesapeake and present-day Virginia. While Algonquians did not claim any form of property over animals—and had domesticated very few—Virginia's governor Francis Wyatt proclaimed in 1623 that anyone convicted of stealing any 'Beast or Bird of Domestical or tame nature' worth more than twelve pence would be put to death.[7] By then colonists' cattle and other domesticated animals were often left to run wild due to the lack of fenced pasture and manpower. These loose animals became a source of recurrent litigation either because they spoiled farmers' crops or were 'stolen.' Preserving the status of animals as property was a way to reinforce the authority of the English rule as wild livestock were considered as property of the Crown.

The notion of animals as property proved essential to the diffusion of animal farming, particularly dairying, as it was used to rationalise the taking of milk from female animals for human consumption.

3   Milk Colonialism

Jonathan Saha has described animal milk as a 'conquering colonial commodity.'[8] The white fluid has indeed been caught up in some of the central tensions of nationalist projects both in the metropoles and their colonies. Before the modern colonisation era, dairying and animal milk consumption were confined to a few regions: central and northern Europe, the Middle East, sub-Saharan Africa, Central Asia, and the Indian subcontinent. To this day, 'lactase persistence' (the ability to digest lactose) remains relatively rare among humans: about 75% of the world population is lactose intolerant. Lactase persistence has been tied to population genetics, which explains why it is found primarily in people with ancestry in territories that have a history of animal domestication such as North-Western

---

[5]Ibid., 80.

[6]Francione, *Animals, Property, and the Law* 1995, 4.

[7]Anderson, *Creatures of Empire* 2004 (note 3), 124.

[8]Saha, 'Milk to Mandalay: Dairy Consumption, Animal History and the Political Geography of Colonial Burma' 2016, 2.

Europe, South Asia, and the Middle East.[9] The fact that animal milk and dairy products are now ubiquitous around the globe, either because they are produced in regions with little or no history of dairying or because they are imported, is a testament to the sway of milk colonialism and international trade law. India, China, Brazil, and New Zealand, all formerly colonised lands, currently figure on the list of the top ten cow's milk producers in the world. The Asia Pacific is one of the biggest markets for imported condensed milk and other processed dairy products.

Deborah Valenze has depicted the global history of milk as the emergence of a culturally malleable, universal commodity, a story 'of [the] conquest of space, energy, and dietary preferences.'[10] The arrival of domesticated ruminants in colonised lands was driven by European agendas. Settlers had become used to consuming milk products back home and were eager to carry on their foodways. Fast forward to the late nineteenth century, dairying had become a major industry in Europe and the United States through economic rationalisation and new technologies which transformed milk from a substance that spoiled so easily that it had to be consumed on the spot into a commodity that could travel huge distances. Condensed milk—still ubiquitous today in formerly colonised countries of East Asia and Africa in particular—is a paradigmatic imperialist food. Its invention by American Gail Borden in the mid nineteenth century was tied to the search for a shelf stable food for soldiers, explorers, and merchants. During the American Civil War, the continental army embraced it as a cheap and transportable source of calories, soon followed by other armed forces, and later by the poor urban classes of European and North America as well as colonisers and colonised people in various regions where fluid animal milk was hard to come by.[11]

The interwar period saw the first concerted attempts to industrialise dairying outside of Europe and North America. China is a case in point. Traditionally a non-dairying culture, it is now the third largest cow's milk producer in the world. The Chinese dairy industry originated in American missionaries and foreign businessmen's efforts to promote milk production and consumption.[12] In the post-colonial era, milk colonialism carried on in most continents under the guise of international law and commerce, which is reflected in the number of international trade disputes pertaining to milk since World War II. Since the 1960s, the expansion of international food aid allowed Europe and the United States to dispose of their milk surpluses, all the while maintaining stable prices at home. For instance, 'Operation Flood,' a program launched in 1970 and financed in part through the sale of European dairy surpluses through the World Food Program transformed India

---

[9]Wiley, *Re-imagining Milk* 2010, 37.

[10]Valenze, *Milk: A Local and Global History* 2011, 3.

[11]More generally, the culturally and legally privileged position of animal milk and other so-called dairy foods in the United States exemplifies a form of domestic milk colonialism whereby milk's promotion reinforces racial and other forms of subordination. Mathilde Cohen, 'Of Milk and the Constitution' 2017a, 115-228.

[12]Sabban, 'The Taste for Milk in Modern China (1865-1937)' 2014, 188-189.

into the world's largest milk producer. Reminiscent of older forms of animal colonialism, this 'white revolution' proceeded in part by replacing Indian bovine breeds with quick fattening, high yield European breeds.

The intense, and increasingly global, dairying and dairy food consumption has had dramatic effects on female mammals and their young. By taking milk from animals and feeding it humans, particularly human babies, dairying severs the nursing relationship twice: between animal mothers and their offspring and between human mothers and their offspring.

4   Breastfeeding Colonialism

In the early twentieth century, lactating animals were conscripted in a colonial reproductive politics aimed at reforming maternity, understood as the embodied experiences of being pregnant, giving birth, and feeding and caring for infants. Since the nineteenth century, low birth rates and high infant mortality (which was due in part to the unsanitary animal milk many infants were fed before 'safer' substitutes became available) had generated anxieties, particularly in the British and French empires. In the metropoles, eugenicist fears developed about national decline and "racial degeneration." In the colonies, the desire for a larger indigenous labour force and army underlied the declared public health goal of fighting 'depopulation' and 'improving' population health. Population growth was seen as a form of power and child rearing became a national duty.

In this highly racialized populationist project, milk turned into a central nationalist and imperialist tool. Indigenous people and animals were stigmatised as inadequate. Native women were accused of lacking maternal instinct and breastfeeding too long, yet producing mediocre milk. Traditional forms of contraception such as postpartum abstinence and long-term breastfeeding were ridiculed or deplored. Indigenous cows were disparaged as producing milk of inferior quality and in insufficient quantities. A colonial doctor, writing in 1936 about French African colonies thus blames low natality rates among the native population on a combination of prolonged breastfeeding, inferior human milk, and scarce and low quality animal milk.[13] Nancy Rose Hunt magisterially described the colonial regulation of breastfeeding in the Belgian Congo as a tool for population increase.[14] In the early twentieth century, making animal milk available to the colony was thought to promote the fertility of women, both white and black. Early weaning and compulsory bottle-feeding were specifically prescribed to black mothers to reduce their milk supply and encourage return to full fertility. Making animal milk available to the colonies was a way to control women's bodies and to produce more bodies—more

---

[13]Cazanove, 'La Question du Lait dans les Colonies Africaines' 1936, 231 (my translation).

[14]Hunt, '"Le Bébé en Brousse": European Women, African Birth Spacing and Colonial Intervention in Breast Feeding in the Belgian Congo' 1988, 401-432.

whites to exploit the labour of black workers, but also more blacks 'to send to Europe the wealth buried in its [Congo's] soil,' as a colonial woman declared in 1926.[15] The Belgian Congo was not an isolated case. 'By the 1920s, bottle-feeding had become part of colonial policies for combating infant mortality in a number of areas, including colonial Malaya, the Belgian Congo, Sudan, French West Africa, Vanuatu and Fiji, and the Philippines,' writes Tehila Sasson.[16] (Though more research is needed to uncover whether this strategy was systematically coupled with an overt anti-breastfeeding and sexual abstinence discourse.)

According to these colonial policies, improving or modernising maternity meant replacing the human breast by cow's milk. This was also true in the metropoles. Milk depots and clinics multiplied in European and American cities, distributing sanitised animal milk at a time when pasteurisation and formula were not widely available. In the colonies, early childhood interventions aimed at challenging indigenous traditions of mothering in the name of civilisation, modernity, and scientific medicine. These policies were prime examples of what Andrea Wiley has termed 'bio-ethno-centrism," i.e., "the interpretation of other people's bodies and behaviour only in relation to those of one's own body and culture, generally with the view that one's own is "better" than the other, or that one's own is "normal" and others are deviant or somehow "abnormal," or "pathological."'[17] Early twentieth century milk depots in the Philippines are illustrative of this dynamic.[18] Through them, the American colonial government hoped to create 'enlightened mothers,' instructed on the 'proper' care of infants, particularly in terms of sanitation and hygiene. The depots dispensed free or subsidised pasteurised cow's milk obtained, whenever possible, by establishing their own dairies populated by cows imported from Australia.

It is now well known that the spread of animal milk, particularly in the form of infant formula, has had deleterious effects on human babies and their mothers, especially in former colonies. However, the harm to animals ushered in by the globalisation of milk consumption is less familiar. Female animals bred and exploited for their milk live a particularly miserable existence, exposed to extreme physiological demands. They are maintained in a quasi-constant state of pregnancy and lactation via forced insemination or other forms of reproductive technologies, only to have their new-borns removed from them so that humans may express their milk. There is a special harm, for both human and non-human mammals in being prevented from nursing, sometimes referred to as 'weaning distress'[19] in the animal behaviour literature.

---

[15]Ibid., 405.

[16]Sasson, 'Milking the Third World? Humanitarianism, Capitalism, and the Moral Economy of the Nestlé Boycott' 2016, 1200.

[17]Wiley, *Re-imagining Milk* 2010, 4.

[18]Roces, 'Filipino Elite Women and Public Health in the American Colonial Era, 1906–1940' 2017, 484.

[19]Weary/Jasper/Hötzel, 'Understanding Weaning Distress' 2008, 25.

Under natural conditions, the weaning process involves a gradual reduction in milk intake, accompanied by increasing social independence from the mother and increasing intake of solid food. By contrast, farm animals are typically weaned abruptly by separating the young from the mother, often within hours of birth. In this system, both baby animals and their mothers show a distinctive distress response when they are separated—both bellow for several days, sometimes weeks with grief.[20] As Sherry Colb has written, '[l]ike other mammals, cow mothers are extremely attached to their new-born babies and want nothing more than to be able to nurse them. The babies feel this way too, and they find comfort, nourishment, and pleasure in nursing on their mothers.'[21] Greta Gaard thus reports the story of a veterinarian called by a farmer because one of his cows was mysteriously dry.[22] The puzzle was soon solved. The cow had given birth to twins. She had brought one new baby to the barn where he was immediately dispatched to the veal crate. But she had hidden the other one in the woods, furtively nursing her whenever she was allowed to pasture.

Though the full implications of milk and breastfeeding colonialism would warrant a much longer discussion, in the limited space available, I will conclude this chapter by floating a proposal for advancing the global animal law agenda.

5   Conclusion: Toward a Trans-Species Right to Breastfeed

While international law has begun to address issues such as endangered species and biodiversity, the welfare of animals, let alone lactating animals, remains unaddressed. Yet, could it be that lactation, because it is common to all mammals, represents a promising starting point for advocating in favour of stronger international animal welfare protection? In other words, one strategy to promote global animal law, both as a research program and as a branch of law, could be to connect it with other international legal initiatives such as women and children's rights. This could be done in an ecofeminist vein, that is, by taking seriously the idea that the oppression of animals and human females is interconnected and mutually reinforcing.

Milk is a quintessentially intersectional issue, cutting across the human/animal divide.[23] It is produced by female mammals of all species, including women. As pioneering ecofeminist Carol Adams likes to point out, all milk from female animals

---

[20]Padilla de la Torre/Briefer/Reader/McElligott, 'Acoustic Analysis of Cattle (Bos taurus) Mother–Offspring Contact Calls From a Source–Filter Theory Perspective' 2015, 58-68 (analyzing cattle vocalizations and finding that both cows and calves produce distinctive calls when they become separated and preceding reunion and nursing).

[21]Colb, '"Never Having Loved at All": An Overlooked Interest that Grounds the Abortion Right' 2016, 952-53.

[22]Gaard, 'Toward a Feminist Postcolonial Milk Studies' 2013, 612.

[23]Cohen, 'Regulating Milk. Women and Cows in France and the United States' 2017b, 469-526.

is 'breast milk.' One avenue to advance the global animal law agenda may be to reorient international breastfeeding advocacy toward promoting the welfare of lactating animals *of all species*, rather than humans only. International human rights lawyers have considered the idea of a woman's right to breastfeed and a child's right to be breastfed for some time. No such rights have been recognised yet, but the World Health Organization (WHO) and the United Nations Children's Fund developed the International Code of Marketing of Breast-Milk Substitutes in 1981 as a health policy framework for breastfeeding promotion. Though the WHO lacks enforcement mechanisms, the Code's recommendations were incorporated into many domestic laws and successfully pressured Nestlé and its like to change their marketing strategies. Why not expand the movement to other lactating animals and their offspring? This shift would vindicate Katsi Cook's beautiful insight that the mother's body is the first environment,[24] and as such should be protected regardless of species.

Acknowledgements For helpful conversations and comments on earlier drafts, I thank Erin Delaney, Amy DiBona, and Elizabeth Emens.

References

Anderson, V. D. (2004). *Creatures of Empire. How domestic animals transformed early America*. New York: Oxford University Press.

Cazanove, J. L. F. (1936). La Question du Lait dans les Colonies Africaines. *Africa, 9*, 227–236.

Cohen, M. (2017a). Of milk and the constitution. *Harvard Journal of Law and Gender, 40*, 115–228.

Cohen, M. (2017b). Regulating Milk. Women and Cows in France and the United States. *American Journal of Comparative Law, 65*, 469–526.

Colb, S. F. (2016). "Never having loved at all": An overlooked interest that grounds the abortion right. *Connecticut Law Review, 48*, 933–967.

Crosby, A. (1986). *Ecological imperialism: The biological expansion of Europe, 900–1900*. Cambridge: CUP.

Francione, G. L. (1995). *Animals, property, and the law*. Philadelphia: Temple University Press.

Gaard, G. (2013). Toward a feminist postcolonial milk studies. *American Quarterly, 65*, 595–618.

Hunt, N. R. (1988). "Le Bébé en Brousse": European Women, African birth spacing and colonial intervention in breast feeding in the Belgian Congo. *International Journal of African Historical Studies, 21*, 401–432.

LaDuke, W. (1999). *All our relations: Native struggles for land and life*. Cambridge: South End Press.

Padilla de la Torre, M., Briefer, E. F., Reader, T., & McElligott, A. G. (2015). Acoustic analysis of Cattle (Bos taurus) mother–offspring contact calls from a source–filter theory perspective. *Applied Animal Behaviour Science, 163*, 58–68.

Roces, M. (2017). Filipino elite women and public health in the American Colonial Era, 1906–1940. *Women's History Review, 26*, 477–502.

---

[24]LaDuke, '*All Our Relations*, 22 (quoting Katsi Cook)' 1999.

Sabban, F. (2014). The taste for milk in modern China (1865–1937). In J. A. Klein & A. Murcott (Eds.), *Food consumption in global perspective* (pp. 182–208). Basingstoke: Palgrave Macmillan.

Saha, J. (2016). Milk to mandalay: Dairy consumption, animal history and the political geography of colonial Burma. *Journal of Historical Geography, 54*, 1–12.

Sasson, T. (2016). Milking the third world? Humanitarianism, capitalism, and the moral economy of the Nestlé Boycott. *American Historical Review, 121*, 1196–1224.

Valenze, D. (2011). *Milk: A local and global history*. New Haven: Yale University Press.

Weary, D. M., Jasper, J., & Hötzel, M. J. (2008). Understanding weaning distress. *Applied Animal Behaviour Science, 110*(2008), 24–41.

Wiley, A. (2010). *Re-imagining milk: Cultural and biological perspectives*. London: Routledge.

Mathilde Cohen is Professor of Law at the University of Connecticut. Her scholarship focuses on various modes of disenfranchisement in French and American legal cultures. Her approach is cross-disciplinary, spanning a variety of subjects, including deliberative democracy, judicial decision-making, and the gendering and racialization of food and body fluids. A graduate of the École Normale Supérieure and the Sorbonne in Paris, she was a research fellow at the French Centre National de la Recherche Scientifique before joining UConn.

Part II
Animals as Commodity

Chapter 5
Trading in Sacrifice

Kristen Stilt

Abstract The chapter examines the trade of live animals for slaughter, focusing on export from Australia to the Muslim-majority countries that are the main customers. Here, animals are shipped across boundaries of religion, culture, and norms of animal welfare. While the typical rules of international trade in goods apply, they do not really fit. In addition, the current legal regime governing live exports is insufficient to provide animals with an adequate standard of welfare, from the point of entering the ships in the country of origin to the moment of slaughter in the importing country. Stilt argues, however, that with the due involvement of religious authorities, the Islamic tradition of animal welfare could be harnessed to develop more widely accepted international transportation and slaughtering standards.

1 Introduction

The international trade of live animals, especially animals sold for slaughter, creates significant challenges for international law. Non-human animals do not fit neatly into the legal world created by humans. In nearly every jurisdiction, animals are property, but they are not like all other property. The sentience of animals has been widely recognized and it forms the basis of anti-cruelty laws where they exist. You may destroy your toaster any way you like, but the laws of most jurisdictions protect how you treat your dog. This fractured point in the law, animals as property and yet not

Revised version of the original published article "Trading in Sacrifice" by Kristen Stilt, American Journal of International Law Unbound, Volume 111, 2017, pp. 397–401. The original article was published as an Open Access article, distributed under the terms of the Creative Commons Attribution licence (http://creativecommons.org/licenses/by/4.0/).

K. Stilt (✉)
Harvard University, Law School, Cambridge, MA, USA
e-mail: kstilt@law.harvard.edu

© The Author(s) 2020
A. Peters (ed.), *Studies in Global Animal Law*, Beiträge zum ausländischen öffentlichen Recht und Völkerrecht 290,
https://doi.org/10.1007/978-3-662-60756-5_5

exactly property, is the source of confusion in national laws, leading to unsatisfactory answers to questions such as what damages should be paid when a companion animal is negligently killed[1] or whether individuals should own wildlife as "pets."

The live animal trade for slaughter adds additional complexities. One of the most pressing issues is how to provide animals an adequate standard of welfare, from the point of entering the ships in the country of origin to the moment of slaughter in the importing country. This is a particularly difficult issue when, as is often the case, the animals are shipped across boundaries of religion, culture, and norms of animal welfare. The typical rules of international trade in goods cannot be sufficient, because animals are simply not like the containers of toasters in international shipping channels. Exporters of toasters do not have expectations for how their products will be treated by purchasers, and when shipments are destroyed in transit, the loss is merely one for insurers to assess.

This essay asserts that current international law is insufficient to provide animals adequate standards of welfare in live export contexts, and it also offers suggestions for improvements. The essay focuses on one of the most significant exporters of live animals for slaughter, Australia, and the Muslim-majority countries that constitute the vast majority of Australia's customers. First, it introduces basic issues in the live export trade, including the concept of *halal* slaughter, which is the means of slaughter in these importing countries. Second, the essay focuses on particular religious reasons for the demand for live animals in Muslim-majority countries, using Saudi Arabia as an example. Third, the essay briefly surveys the current legal regime governing live exports, identifies its flaws, and makes recommendations. Understanding religious beliefs and motivations is essential because devising an international legal regime that adequately protects the welfare of animals must take into account the fact that the animals are crossing boundaries of religion, culture, and tradition, and in particular conceptually crossing from "west" to "east," which brings its own additional sensitivities and challenges.

2   The Live Export Trade

The requirements for meat to be considered *halal* are relatively few, and slaughterhouses around the world have been certified as *halal* by many different Muslim organizations. For most farmed animals, the method typically involves cutting the animal's carotid arteries, jugular veins, trachea, and esophagus with a sharp instrument such that the animal dies of blood loss, or exsanguination. The name of God must be mentioned at the time of slaughter, and the slaughterer should be a Muslim, Christian, or a Jew.[2]

---

[1] Supreme Court of Texas, *Strickland v. Medlen*, 2013, 397 S.W.3d 184.

[2] Usmani, *Islamic Laws* 2006, 25-50.

As one of the world's largest producers of farmed animals, Australia is home to many *halal* slaughterhouses, and the country ships frozen *halal* meat globally. Yet millions of animals each year also are shipped, for *halal* slaughter, from Australia to Middle Eastern and Asian countries that cannot provide enough domestic supply to meet consumption demands. The reasons for a preference for locally slaughtered meat in these importing countries are complex, and include differing views on *halal* standards, and in particular a rejection of the pre-slaughter stunning that is required in *halal* slaughterhouses in Australia; support for the local feedlot and meat processing industry; and the lack of refrigerated facilities in some areas. The reasons also include the desire by some Muslims to slaughter an animal on certain occasions and on Eid al-Adha in particular, the holiday commemorating Abraham's willingness to sacrifice his son. Before Abraham could do so, God sent an animal as a substitute sacrifice.

Cattle, sheep, and goats raised in Australian pastures, grazing freely, do not always transition well to transport ships in which tens of thousands of animals are housed in small spaces for journeys up to 1 month. The animals do not always manage to eat the pelleted feed on the ships, even after a transitional period in a feedlot, and they are often travelling from the Australian winter to the Middle Eastern summer, which can lead to heat stroke. Mortality rates of 2% for sheep and 1% for cattle are considered normal. Higher death rates require the exporter to notify the Australian authorities, who may, but are not required, to conduct an investigation. This means that of 70,000 sheep, the death of less than 1400 is considered an acceptable loss of inventory. When a problem in transport arises, the results can be catastrophic. The Australian authorities can prevent a shipment from returning to Australia due to biosecurity reasons. In September 2012, for example, 20,000 Australian sheep were sent to Bahrain, but were rejected prior to unloading on allegations of suspected disease. Eventually, Pakistan accepted the shipment. But when it learned about the Bahraini rejection, the Pakistani authorities killed the sheep in brutal ways.[3]

Other challenges to animal welfare occur upon arrival at the destination country. The animals are subject to local policies and laws, which most likely do not include an animal welfare law. It is also unlikely that pre-slaughter stunning is practiced in the receiving country. The issue of stunning is a key area of contention in the world of international *halal*, creating two camps: those who believe that pre-slaughter stunning reduces an animal's pain and suffering and does not jeopardize the *halal* status of the meat because the stunning does not cause the animal's death, and those who argue that stunning was never a part of traditional Islamic slaughter and causes harm to the animal. Those in the latter group also express concern that stunning could kill the animal, even though it is not the intent, thus running the risk that the meat is *haram*, or impermissible for human consumption, because the animal did not die from exsanguination.

---

[3]Four Corners, 'Another Bloody Business', 5 November 2012, available at: http://www.abc.net.au/4corners/another-bloody-business/4354700.

3 Religious Motivations for the Importation of Live Animals

Saudi Arabia provides an example of the demand for live animals for *halal* slaughter and in particular for sacrifice during the annual pilgrimage, or hajj. Saudi Arabia has not imported Australian sheep since 2012 when it rejected Australia's new regulations, as discussed below, but talks are underway in an effort to resume the trade and in the meantime Saudi Arabia is sourcing its animals from other countries. The hajj example is also particularly revealing because it allows for a glimpse at how customary practices and beliefs surrounding *halal* slaughter can change.

The situation with the sacrifice of animals during the pilgrimage, as explained by the Saudi Project for the Utilization of Sacrificial Animals, a state agency, begins with the pilgrimage's obligations:

> Hajj (Pilgrimage) involves certain religious rites that should be performed at particular times and at certain places. Among these rites is the sacrifice of whatever cattle or sheep the pilgrim can easily afford. Pilgrims of diverse races, customs and social levels are anxious to follow the Sunnah of the Prophet (peace be upon him) by slaughtering the sacrificial animals by themselves.[4]

The practice of pilgrims purchasing and sacrificing their own animals became problematic with the increase in the numbers of people making the pilgrimage and with the availability of air transportation in particular. Larger numbers of pilgrims could arrive and could also depart soon after the pilgrimage ended. During their short time in Saudi, they could consume or distribute to others only a limited amount of meat from the animals slaughtered as part of the pilgrimage. The increase in humans making the pilgrimage resulted in an increase in demand for animals for slaughter, leading to even greater excesses of meat. The Saudi Project noted that some pilgrims purchase animals who are not fit for use in sacrifice, while "the majority of pilgrims leave the slaughtered animals at their place of slaughter, not making use of their meat because of overcrowding and the extremely hot weather.[5]

In the early 1980s, the problem of rotting carcasses reached proportions that the Kingdom deemed unacceptable, and so it initiated the Saudi Project for the Utilization of Sacrificial Animals. Managed by the Islamic Development Bank, the Saudi Project "acts as an agent to pilgrims in carrying out on their behalf the slaughtering of all types of offerings they make and the transport and distribution of the meat, all in accordance with the principles of Sharia."[6] Pilgrims may purchase a coupon that authorizes the Saudi Project to slaughter an animal on behalf of the purchaser, or pilgrims may form a group of thirty or more and one person among them will receive a permit granting access to the slaughterhouse as the group's representative. The

---

[4]The Saudi Project for the Utilization of Sacrificial Animals, available at: http://www.isdb.org/irj/ go/km/docs/documents/IDBDevelopments/Internet/English/IDB/CM/ADAHI/ AboutADAHIProject.html.

[5]Ibid.

[6]Ibid.

slaughterhouses are now equipped with high-tech slaughter equipment and operate continuously for 4 days during the pilgrimage, slaughtering up to a million animals.[7] The animals' throats are cut, per the *halal* slaughter practiced in Saudi Arabia, while they are fully conscious—there is no pre-slaughter stunning. These slaughterhouses then preserve the expected excess, which is exported to destinations throughout the Muslim world.[8]

The Saudi Project has attempted to convince pilgrims to utilize the coupon system, rather than slaughtering their own animals as has been the tradition, through a variety of means. The Project provides a *fatwa*, or legal opinion, addressing the issue of the timing of the slaughter. This issue arises because when a pilgrim purchases a coupon, the Project is not able to tell the pilgrim exactly when the animal will be slaughtered, and activities performed during the hajj are expected to take place in a particular order. The Saudi Project cites a *hadith*, or a saying of the Prophet, in which the Prophet is asked about the permissibility of shaving one's head prior to the sacrifice of an animal and of sacrificing an animal prior to the throwing of stones. To these and other questions about the appropriate sequence of events the Prophet expresses no concern: "no harm, just go and do what is required to be done."[9]

The coupon system, the permissibility of which does not seem to be questioned by the Muslims participating in it, may have solved the problem of rotting carcasses, but it both enables and ignores another. Newer and faster slaughterhouses, and greater numbers of butchers brought to Saudi Arabia from around the Muslim world, now result in the slaughter of nearly a million animals during the pilgrimage alone. In addition to the sacrifice of these animals in Saudi Arabia during the hajj, Muslims around the world also commemorate the occasion with the same practice; estimates of the number of animals slaughtered worldwide on Eid al-Adha are impossible to make. Beyond Eid al-Adha, of course, is the need for animals for daily consumption, all of which contributes to the demand for live exports.

---

[7]In 2010, al-Jazeera produced a short report on the infrastructural challenges of the slaughter during the hajj. Al-Jazeera, 'The Hajj 2010: Streamlining Ritual Slaughter of Animals', 30 December 2010, available at: https://www.youtube.com/watch?v=85RXMB1Hauc. In 2014, Saudi imported over 1 million animals and a total of 2.5 million were sold throughout the country for Eid al-Adha. Syeda Amtul, '2.5 m Animals Sold in Saudi Arabia During Hajj', *Al-Arabiya*, 12 October 2014, available at: http://english.alarabiya.net/en/News/middle-east/2014/10/12/2-5m-animals-sold-in-Saudi-Arabia-during-hajj.html.

[8]'The Saudi Project for the Utilization of Sacrificial Animals', available at: http://www.isdb.org/irj/go/km/docs/documents/IDBDevelopments/Internet/English/IDB/CM/ADAHI/AboutADAHIProject.html. In recent years, outside of Saudi Arabia, Bangladesh has received the largest amount of frozen meat.

[9]Ibid.

4   The Legal Regime Governing Live Exports

Turning to the legal regime that governs Australia's exports, prior to 2011, the system was relatively simple. Australian law regulates exports and requires the exporter to hold a livestock export license and permit and provide a Notice of Intention for Export. Exporters must comply with the Australian Standards for the Exports of Livestock, which provide animal health and welfare requirements, such as stocking density, during transportation and up to the point of disembarkation. The exported livestock must also meet importing country requirements. In 2011, an exposé changed Australia's live export industry. Investigations conducted in Indonesia by the NGO Animals Australia led to Australia's leading investigative news program, Four Corners, producing an hour-long investigative segment about the Indonesian slaughterhouses that received Australian cattle.[10]

The Four Corners program was the most significant media coverage that the live export issue had received, and it shocked the Australian public. Four Corners investigators revealed images from Indonesian slaughterhouses showing workers kicking and beating the cattle, breaking their tails, and using abusive tactics to bring the massive animals to the ground for slaughter. The throats of fully conscious cattle were cut, in one case up to 33 times. Farmed animal expert Temple Grandin commented that "the conditions are absolutely terrible." Based on public outrage, the Australian government suspended cattle exports to Indonesia for a month.

A new scheme was quickly implemented that requires exporters to have an approved Exporter Supply Chain Assurance System (ESCAS) before receiving an export permit. ESCAS requires the exporter to control the animals' care and handling all the way to the point of slaughter and ensure that the handling at all stages complies with the OIE (which is also known as the World Organisation for Animal Health) Terrestrial Animal Health Code. This system is intended to extend protection over the animal throughout the entire "supply chain." Importantly, this is not Australian law that travels with the animals but rather the much weaker OIE standards. The OIE has historically focused on animal diseases, and has only recently included animal welfare, as relates to farmed animals, in its scope of concern.[11] The Code provides very minimal recommendations that all 181 countries that have joined the OIE are expected to be able to meet, although the standards are non-binding unless the member country adopts them through national legislation. For example, Article 7.5.2 includes the provision that "Animals for slaughter should not be forced to walk over the top of other animals." Australian law requires pre-slaughter stunning while the OIE Code does not. Further, as a practical matter, the exporter cannot always control what happens to the animals once they arrive in their destination countries even though they are required to do so by ESCAS.

---

[10]Four Corners, 'A Bloody Business', 30 May 2011, available at: http://www.abc.net.au/4corners/a-bloody-business%2D%2D-2011/2841918.

[11]Shortcomings of the OIE and the Codes that it has produced are discussed in Peters, 'Global Animal Law' 2016.

Most recently, the conditions on board the transportation ships for sheep were exposed through undercover video footage taken by a member of the crew and presented on the Australian news program 60 Minutes.[12] On one voyage the crewmember sailed in August 2017, about 2400 sheep, more than 3% of those onboard, died due to the heat. The sheep, packed into pens on the multi-story ships, literally cooked to death, "covered in waste and desperately gasping for air and water in extreme heat."[13] The bodies of the dead sheep were thrown overboard into the sea. The exposé prompted calls within Australia for a ban on the live export trade.[14]

These exposés and the resulting reactions in Australia shows that Australians want better treatment for the animals they send for slaughter, and recipient countries' animal advocates, small in number in comparison to the Australian animal protection movement but vocal and growing, also want better treatment.[15] In the absence of a meaningful international framework, Australia, as with other live exporting countries, has the discretion to determine what standards it imposes on its exporters. The Australian industry, and government, is concerned that it will lose access to markets to other exporters, such as Brazil, if its demands are too high, and especially if stunning is required, and so it has not imposed it as a condition.

A successful international legal system that provides meaningful protections to animals in the live export trade must do more than just impose new standards, especially when religious beliefs stand behind not only the demand for live animals but also, in some cases, the kind of *halal* slaughter in the recipient country. The Saudi hajj example is instructive. Despite a longstanding tradition, and indeed a religious belief on the part of pilgrims that the individuals needed to slaughter their own animals, Saudi Arabia changed the process and backed it with religious arguments. The new system probably results in the slaughter of more animals overall, under governmental control, but the point remains that religious beliefs can change and credible religious authorities can accelerate that change. Many Muslim religious leaders worldwide have approved the use of pre-slaughter stunning and it is now widely used in Jordan and Indonesia. Involving these actors as allies can help to generate norms that are "particularly sensitive to problems of Eurocentrism, of legal imperialism, and of a North-South divide."[16]

More generally, there is a deep tradition of animal welfare in Islamic law, which animal advocates in the Muslim world are working hard to revive and even

---

[12]60 Minutes Australia, 'Sheep, Ships and Videotape', 8 April 2018, available at: https://www.youtube.com/watch?v=m1V96Y533Ds (Part 1) and https://www.youtube.com/watch?v=FR09We_f9U4&t=2s (Part 2).

[13]Jane Dalton, 'Australian sheep bound for the Middle East "cooked alive in deadly and harrowing conditions" on live export ships', *Independent* (9 April 2018).

[14]Ibid.

[15]See Stilt, 'Constitutional Innovation' 2018.

[16]Peters, 'Global Animal Law' 2016, 22.

expand.[17] There are Muslim advocates in the importing countries who question whether live exports can ever comply with Islamic notions of animal welfare.[18] Further, the recent exposé of sheep transportation from Australia to the Middle East raises the possibility that the *halal* status of the meat from the animals who do survive the journey could be in jeopardy.[19] The ongoing efforts by scholars and advocates to recover the strong tradition of animal welfare in Islamic law and interpret it in light of contemporary circumstances show that more effective and widely accepted international standards are politically and practically feasible.

References

Peters, A. (2016). Global animal law: What it is and why we need it. *Transnational Environmental Law, 5,* 9–23.
Stilt, K. (2017). Animals. In A. Emon & R. Ahmed (Eds.), *The Oxford Handbook of Islamic law.* Oxford: Oxford University Press.
Stilt, K. (2018). Constitutional innovation and animal protection in Egypt. *Law & Social Inquiry, 43,* 1364–1390.
Usmani, M. T. (2006). *The Islamic laws of animal slaughter* (pp. 25–50). Santa Barbara: White Thread Press.

Kristen Stilt is a Professor of Law at Harvard Law School. She also serves as faculty director of the Animal Law & Policy Program, faculty director of the Program on Law and Society in the Muslim World, and is a deputy dean. Stilt was named a Carnegie Scholar for her work on Constitutional Islam, and in 2013 she was awarded a John Simon Guggenheim Memorial Foundation Fellowship. Her research focuses on animal law (comparative, international, and religious law); Islamic law and society; and comparative constitutional law.

---

[17]Stilt, 'Animals' 2017; Stilt, 'Constitutional Innovation' 2018.

[18]In 2006, the Egyptian animal advocates sought a legal opinion, or *fatwa*, on the issue of live exports from Muhammad Tantawy, who at that time was the Shaikh of al-Azhar, one of the two highest religious positions in the country. Ahmed Sherbiny, President of the Egyptian Society of Animal Friends, submitted the question, which described the conditions during long distance transport from Australia to Egypt and asked for the Islamic legal status of such transport. In his reply, Sheikh Tantawy stated, "Causing pain to the animal during transport as described in the letter is considered an action prohibited and forbidden in Islamic law, assuming, of course, that the situation is as described in the letter." Fatwa on file with author.

[19]Victoria Laurie, 'Halal Doubts Cast on New Sheep-Shipping Methods', *The Australian* (21 May 2018).

Chapter 6
Cross-Border Forms of Animal Use by Indigenous Peoples

Stefan Kirchner

Abstract The chapter discusses animal use by indigenous peoples that involve crossing state borders, using the example of reindeer herding by indigenous Sámi in Sweden, Norway, and Finland. Animals play important cultural, economic, and spiritual roles for indigenous communities. This particular form of interaction between humans and animals is, however, not sufficiently recognized by contemporary laws. The risk of overruling the interests of migratory animals, and of the pastoralist (semi-)nomadic human communities depending on them, is exacerbated when the herds cross boundaries.

1 Introduction

The aim of this essay is to show how international law relates to the interaction of indigenous peoples and animals across international borders. While colonial borders have affected the lives of herding communities in Africa and while there are cross-border indigenous activities in different parts of Latin America, the situation in Northern Europe is particularly noteworthy. This is because many kinds of cross-

Revised version of the original published article "Cross-Border Forms of Animal Use by Indigenous Peoples" by Stefan Kirchner, American Journal of International Law Unbound Volume 111, 2017, pp. 402–407. The original article was published as an Open Access article, distributed under the terms of the Creative Commons Attribution licence (http://creativecommons.org/licenses/by/4.0/).

The author thanks Sámi friends and colleagues from Sweden, Norway and Finland for their insightful information about Sámi livelihoods. All opinions and errors contained in this text are only attributable to the author.

S. Kirchner (✉)
University of Lapland, Arctic Center, Rovaniemi, Finland
e-mail: stefan.kirchner@ulapland.fi

© The Author(s) 2020
A. Peters (ed.), *Studies in Global Animal Law*, Beiträge zum ausländischen öffentlichen Recht und Völkerrecht 290,
https://doi.org/10.1007/978-3-662-60756-5_6

border activities are possible there, not simply because effective border controls are difficult to ensure in such remote areas, but mainly because several of the relevant states have the long-term political will to allow for cross-border activities. This freedom is not enjoyed by indigenous communities who have long called the area home. Particular attention will therefore be given to the situation of the indigenous Sámi people. Their homeland, Sápmi, is ruled by Norway, Sweden, Finland and Russia.

Home to the indigenous Sámi people for thousands of years, the region eventually saw the arrival of Germanic-speaking peoples from the south.[1] For centuries, overwhelmingly large parts of Fennoscandia, with a few exceptions, were dominated by non-indigenous powers. The 1751 Peace Treaty of Strömstad created a border between Denmark (which at the time included Norway) and Sweden (which then also included Finland) included a supplement, the so called *Lapp Codicil*. This document guaranteed the continued practice of migratory[2] Sámi reindeer herding in the region and therefore also across the newly created international border,[3] as well as other traditional ways of livelihood, such as fishing.[4] The practical reality in over the coming centuries, though, looked very different[5] and the borders between Russia (which ruled Finland between 1809 and 1917) and Norway and Sweden were closed in 1852 and 1889 respectively.[6] Russian rule over Finland also led to the effective end of representational rule, which also led to an end to the legal recognition of the Sámi community, the *siida*,[7] as a relevant political entity, even though it continued to remain relevant in the remote Northwestern part of Finland.[8] The creation of national legislation on reindeer herding has moved the decision-making on reindeer herding[9] from the traditional model of *siida* reindeer herding communities to the national administration and the reindeer herding associations.[10] Also in the

---

[1]Broadbent, *Lapps and Labyrinths* 2010, ebook, position 4137. Note that, while still found in place names, the use of the term "Lapp" to refer to Sámi people is today considered offensive, Fuglerud, 'Positioned Creativity' 2016, 162. See also Brännlund/Axelsson, 'Reindeer management' 2011, 1097 et seq.

[2]The term 'migratory reindeer herding' refers to the practice to moving reindeer between summer and winter pastures (see Næss et al., 'Cooperative pastoral production' 2010, 249) which is still practiced in Sweden and Norway, but not in Finland.

[3]Heikkilä, *Reindeer Talk* 2006, 102.

[4]See Kent, *Sámi Peoples* 2014, 240.

[5]See Lantto, Borders, Citizenship and Change 2010.

[6]Heikkilä, *Reindeer Talk* 2006, 103. On the impact of borders on the Sámi people see also Seurujärvi-Kari/Carpelan, 'Regions and borders' 2005, 294-295 and Lehtola, *Sámi People* 2010, 78-79.

[7]See Kent, *Sámi Peoples* 2014, 228-229.

[8]Heikkilä, *Reindeer Talk* 2006, 113.

[9]For an overview over the situation in Norway and Finland see Heikkilä, *Reindeer Talk* 2006, 115-116.

[10]Heikkilä, *Reindeer Talk* 2006, 114. On the different levels of autonomy of reindeer herding communities in Norway, Sweden (in both countries only Sámi) and Finland (Sámi and non-Sámi) see Reinert, 'Economics of reindeer herding 2006, 537.

other parts of Sápmi, the states play a dominant role in shaping and regulating reindeer herding and even in the areas which are recognized by the states as Sámi homelands, "reindeer herding has lost its primary land use status in favor of agriculture, forestry, mining industry, water power construction or tourism".[11] This trend is only going to become more relevant as climate change is making large parts of the Arctic and Sub-Arctic more accessible to outside economic interests.

Today, cross-border reindeer herding in Sápmi is limited to the border between Norway and Sweden, while there is no cross-border reindeer herding involving Finland or Russia. European integration has not changed this situation: the borders between Norway, Sweden and Finland have been open since the Nordic Passport Union of 1952, significantly predating the Schengen Agreement of 1985, which allows for unhindered travel in large parts of Europe, including these countries.[12] Finland and Sweden are members of the European Union, while Norway is part of the European Free Trade Area and of the Schengen Agreement, Russia imposes visa requirements on citizens of the three other states. With such limitations, the Russian part of Sápmi is effectively cut off from the Western parts. While the borders between Finland, Norway and Sweden have long been open for many purposes, this openness does not fully take into account the needs of the indigenous Sámi people, who consider themselves to be one people and consider the Sápmi homeland as a whole.[13] This is reflected in ongoing disputes concerning the legal basis for cross-border reindeer herding between Norway and Sweden. Today, only part of their ancestral homeland is recognized as Sámi home areas in the legal sense of the term and the Sámi are a minority in their own regions virtually throughout Sápmi.[14]

---

[11]Heikkilä, *Reindeer Talk* 2006, 115.

[12]The Schengen *acquis* as referred to in Article 1(2) of Council Decision 1999/435/EC of 20 May 1999, Agreement between the Governments of the States of the Benelux Economic Union, the Federal Republic of Germany and the French Republic on the gradual abolition of checks at their common borders, Official Journal 2000 L 239, available at http://eur-lex.europa.eu/legal-content/ EN/TXT/PDF/?uri=OJ:JOL_2000_239_R_0001_01&from=EN, 22 September 2000, pp. 13-18, for the implementation of the Schengen agreement see the Convention implementing the Schengen Agreement of 14 June 1985 between the Governments of the States of the Benelux Economic Union, the Federal Republic of Germany and the French Republic on the gradual abolition of checks at their common borders, ibid., pp. 19-26; for the accession to the latter by Finland and Sweden see ibid., pp. 106-114 and pp. 115-123, respectively.

[13]*Cf.* United Nations Regional Information Centre for Western Europe, *The Sami of Northern Europe – one people, four countries*, available at http://www.unric.org/en/indigenous-people/ 27307-the-sami-of-northern-europe%2D%2Done-people-four-countries.

[14]Also the part of Finland in which reindeer herding is conducted is significantly larger than the Sámi home area. For a map of the former see Åsbakk/Kumpula/Oksanen/Laaksonen, 'Infestation' 2014, 173, for a definition for the latter see Section 4 of the Finnish Act on the Sámi Parliament, 974/1995, as amended, unofficial translation by the Ministry of Justice, Finland, available at https:// www.finlex.fi/en/laki/kaannokset/1995/en19950974.pdf.

2 Traditional Livelihoods of a Transnational Indigenous People

Reindeer herding is an important source of income for many indigenous communities in the Arctic and Sub-Arctic,[15] especially in the Nordic countries and Russia. Reindeer are owned but for almost all practical purposes are free to roam large, but delimited, areas. In Norway and Sweden the human-reindeer interaction involves seasonal migration between summer pastures at higher altitudes and winter pastures closer to the sea.

In Norway and Sweden only indigenous persons have the right to herd reindeer, whereas every EU citizen, enjoying the freedom to work across borders, is free to own and herd reindeer in Finland. In Sweden, the right to herd reindeer is held by the local *sameby*, the Sámi Village. While serving as a reminder of the traditional Sámi unit of organization, the *siida*, these entities, despite their name, have been created by the state for the purpose of managing the right to herd reindeer. Each *sameby* holds the right to herd reindeer in a long thin strip of land, generally ranging from the mountains in the West to the sea in the East. While there are still many parallels to traditional Sámi forms of organizing and managing reindeer herding, the framework for reindeer herding is controlled by the state rather than by indigenous communities. Today, reindeer herding is a business like many others and the state's view on reindeer herding can be summarized to the effect that "reindeer herding is about managing privately owned capital (the reindeer) on a common resource base (the pasture)".[16] Land use conflicts between reindeer herding and other land uses, such as forestry,[17] and the decreasing availability of land suitable for reindeer herding put significant economic stresses on Sámi reindeer herders, which come on top of decades long declines in the incomes of Sámi reindeer herders.[18]

In Finland, indigenous reindeer herding families are even more constrained by the legal framework created by the state: reindeer herding is organized through reindeer herding associations (*paliskunnat*).[19] Within a *paliskunta*, decisions are made by the majority, but because non-indigenous persons may herd reindeer, too, it is possible for Sámi reindeer herders to be in the minority. As a result, the traditional ways of herding reindeer have been abandoned in many *paliskunnat*. In one *paliskunta*, the non-indigenous majority sets the numbers of reindeer to be slaughtered, which effectively forced indigenous reindeer herding families in one part of the *paliskunta* to slaughter all of their reindeer.[20]

As a result of the national regulation of reindeer herding, today there is no longer any transboundary reindeer herding in Sápmi: while the free movement of persons is

---

[15]Stoyanova. 'Saami facing the impacts' 2013, 291.

[16]Heikkilä, *Reindeer Talk* 2006, 133 et seq., italics omitted.

[17]Cf. Roturier/Roué, 'Of Forest, Snow and Lichen' 2009, 1960.

[18]Cf. Reinert, 'Economics of Reindeer Herding' 2006, 523.

[19]See in detail Mustonen/Jones, *Reindeer Herding in Finland* 2015, 4 et seq.

[20]Human Rights Committee, *Kalevi Paadar et al. v. Finland*, Communication No. 2102/2011 of 10 April 2014, U.N. Doc. CCPR/C/110/D/2102/2011 (2014).

permitted under the Schengen Agreement and many people in the region cross borders on a regular basis, reindeer herding is regulated on the national levels. This has long been a problem for Sámi reindeer herders.[21]

This indicates a stronger role for the nation state in a region which has long been defined by shared cultures rather than borders. Centuries ago, the Sámi were taxed by different sovereigns. Today, although their way of life has not been entirely robbed of its transnational dimension, their best-known traditional livelihood has been restricted by borders that were created without the doing of the indigenous communities affected. Despite the fact that most Sámi depend primarily on other sources of income, reindeer herding remains an important part of Sámi culture.

3   Involving Indigenous Peoples in Decision-Making Processes and Protecting Their Rights in International Agreements

One important problem of indigenous rights in the Arctic is the involvement of indigenous communities in decision-making processes. Today, national Sámi Parliaments in Norway, Sweden and Finland play important roles in giving the Sámi people a voice on the national level, for example, in the context of mining permits. However, the rights of indigenous peoples are at risk of being ignored by national governments engaging in international relations.

For example, the lack of Sámi participation was criticized recently in the context of the Deatnu River Agreement between Finland and Norway. The Deatnu River (Finnish: Tenojoki, Norwegian: Tanaelva) forms part of the border between Finland and Norway in the homeland of the Sámi people. Next to reindeer herding, fishing is a key traditional livelihood of the Sámi people. The Deatnu River is one of the best salmon rivers worldwide and attracts tourists from around the world due to its remarkable water quality and the size of the salmon caught there. This sets the stage for a usage conflict between indigenous and non-indigenous actors. The agreement between Finland and Norway restricts traditional indigenous fishing rights in the Sámi heartland, which *de facto* allows for the development of fishing-based tourism in the region. While fishing and tourism are important sources of income for this region of Europe, fishing is a crucial element of Sámi culture as well. This is particularly so not only along the coasts but especially in the North-Eastern part of Finland and in Norway.

By agreeing on the distribution of the right to fish in the border river without taking sufficient account of the local indigenous communities,[22] Finland and

---

[21] See already Elbo, 'Norwegian-Finnish reindeer fence' 1954, 73-74.

[22] For an overview over the fishing regulations under the agreement between Finland and Norway see Ministry of Agriculture and Forestry Centre for Economic Development, Transport and the Environment/Fisheries Division LUKE/Natural Resources institute Finland/ River Tenojoki

Norway risk violating the rights of the indigenous Sámi to be heard, their right to be consulted (under ILO Convention No. 169 for Norway and, arguably more far-reaching, under customary international law for Finland, which has not yet ratified the Convention), and their right to engage in sustainable livelihoods.[23]

One notable exception to the lack of Sámi participation is the Sámi Parliaments' role in drafting the proposed 'Nordic Saami Convention' ('Saami' is an alternative spelling for 'Sámi'). The need for a regional treaty arose due to the limited of protection of indigenous rights under existing norms, as the next paragraphs will explain.

The transnational reality of Sámi identity is not recognized in international law, a problem this group shares with many other transnational indigenous peoples. In general, unlike in the Inter-American human rights system, indigenous rights have hardly played a role at the European Court of Human Rights.[24] The European Convention on Human Rights (ECHR), which has been ratified by the four states in question, only protects indigenous livelihoods in the context of the right to private life (Article 8 of the ECHR).[25]

Some Sámi litigants have sought protection of their rights under Article 27 of the International Covenant on Civil and Political Rights (ICCPR), also ratified by the four states, and some work has been undertaken in Europe on national minorities. Article 27 of the ICCPR remains a norm of central importance in the context of indigenous rights, especially as ILO Convention No. 169 has not been ratified by key states in the region, including Sweden, Finland and Russia, and the 2007 UN Declaration on the Rights of Indigenous Peoples (which in large parts reflects the existing customary international law) is non-binding as it was adopted by the UN General Assembly. The importance of Article 27 of the ICCPR is reflected in the range of cases in which it has been invoked, many of which originated from Sápmi. According to Article 27, minorities, including indigenous communities, have the right to enjoy their culture. This has long been understood to include cultural activities which have the purpose of making a living, including reindeer herding for profit. The wording of Article 27 of the ICCPR, which dates back to the 1960s, however, shows that existing international human rights norms are still too dependent on the current state-centered system of international law in order to fully take into account the needs of transnational indigenous peoples: the rights under Article 27 of the ICCPR only apply "[i]n those States, in which ethnic, religious or linguistic minorities exist".[26] There is no transnational element in Article 27 of the ICCPR which would allow for cross-border reindeer herding.

Fisheries Research, *Information about Teno for Fishing Enthusiasts* (2017), available at https://www.ely-keskus.fi/documents/10191/23117928/Teno+info+English.pdf/1a185302-fa5d-4ba6-bd90-17183fefdf21.

[23]On the negative consequences of outside stressors under traditional and modern reindeer herding models see Burkhard/Müller, 'Case study northern Fenno-Scandinavian reindeer herding' 2008, 829 et seq.

[24]But see e.g. Koivurova, 'Jurisprudence' 2011.

[25]See European Commission of Human Rights, *G. and E. v. Norway*, Applications Nos 9278/81 and 9415/81, Decision of 3 October 1983.

[26]Art. 27 ICCPR.

The United Nations Declaration on the Rights of Indigenous Peoples (UNDRIP) could have been an important step towards closing this gap—and the development of customary international law in the years since then is not to be underestimated—but the UNDRIP was adopted 'only' by the General Assembly of the United Nations. It therefore is not a binding document *per se*. While UNDRIP is contributing to the development of customary international law, codifications, especially 'user-friendly' codifications, are a hallmark of the legal systems of the Nordic states. It is therefore hardly surprising that there has long been a desire to clarify in a written instrument the legal situation of the Sámi people across borders.

Today, even the definition of who is actually Sámi varies between the states because the states have set different criteria for the eligibility to participate in the elections to the respective national Sámi Parliaments. In recent years, negotiations have been underway between Norway, Finland and Sweden to establish an international treaty, the Nordic Saami Convention, to unify the legal position of the Sámi people at least in three of the four countries in question. The national Sámi Parliaments played an important role in the drafting process, thereby expanding the realm of consultation of indigenous communities from the national to the transnational—in accordance with the cultural reality of the Sámi people, who consider themselves to be one people, regardless of languages and passports.

The attempt at involving indigenous representatives in the making of a new international treaty is a laudable development, but due to the long passage of time during which reindeer herding has already been regulated on the national level and within national borders, this move might come too late for Sámi reindeer herders. At this time, a revival of old—sustainable—patterns of reindeer herding, regardless of national borders, appears if not impossible then at least highly unlikely. The revival of some Sámi languages in recent years has been an important cultural development, but the recognition of traditional ways of reindeer herding appears to face too many obstacles, not least in the form of competing land use claims, to be a realistic option in the foreseeable future.

In the case of the fishing rights in the Deatnu river it would have been much easier for the two nations, which have long enjoyed excellent relations, to come to a solution which would have allowed the realization of transnational animal use rights by the Sámi people living on both sides of the river. This opportunity was not only missed when the economic interests of the nation states areas were given precedence over indigenous needs and traditions, but also when both states failed to let the affected Sámi people weigh in sufficiently.

4 Animals, Welfare, Animal Welfare and Indigenous Culture

The failure of states to adequately involve transnational indigenous communities in decision-making processes can also have an impact on the welfare of the animals in question. This is particularly the case when the state imposes ways of handling animals which are at odds with methods that have long been known to be sustainable and holistic, in the sense that they might be beneficial for the overall herd and the community. This, however, does not mean that indigenous approaches are always best from the perspective of the welfare of the individual animal. The approach taken by indigenous communities that depend on animals for their survival is usually dominated by pragmatism rather than contemporary animal welfare discussions. Reindeer herding remains an important element of Sámi culture even though today many indigenous persons have other sources of income. Not only is reindeer meat a regionally important source of food, many traditional products are made from reindeer parts—such as antlers or hide.

That said, often the way that Sámi communities treat animals is significantly closer to the needs of animals than factory-based meat production would allow. For example, unlike caribou in North America,[27] reindeer in Sápmi are no longer hunted[28] but herded.[29] The animals are owned by the herders but not fully domesticated in the same way as farm animals are. Far from it, they roam large areas fairly freely and are only collected in autumn[30] or, in Sweden and Norway, during the seasonal migrations between summer and winter pastures.[31] For Sámi reindeer herders, the wellbeing of the animals is essential for their own wellbeing. Indeed, meat production is only one purpose of reindeer herding; other parts of the animals are used for the production of culturally relevant products. The cultural, not merely practical, importance of caring for reindeer differentiates this form of animal use from other animal-related livelihoods, including ranching, hunting or fishing. Indeed, economic profit was never a main goal of reindeer herding. In addition to providing food security, the cultural dimension of hunting, herding and fishing is not to be underestimated. Often key animals take on a highly symbolic value for indigenous cultures. Especially for (semi-)nomadic indigenous communities, as commonly found in the Arctic and Sub-Arctic, gatherings are associated with traditional forms of animal-related livelihoods. For transboundary (semi-)nomadic indigenous peoples like the Sámi, such gatherings are also important cultural events. By reorganizing the traditional cultural activity of reindeer herding within different national legal frameworks, not only the traditional way of life but also the culture of the indigenous Sámi people has been affected.

---

[27]See Alaska Department of Fish and Game, *Caribou Hunting in Alaska*, available at: http://www.adfg.alaska.gov/index.cfm?adfg=caribouhunting.main.

[28]On historic Sámi reindeer hunting see Snatic, *Hunting and Gathering*, available at: https://www.laits.utexas.edu/sami/diehtu/siida/hunting/jonsa.htm. Elsewhere, hunting and poaching remain a serious threat to the survival of reindeer, for example on Russia's Taymyr peninsula, see (no author names), 'How poaching is "killing off" the world's largest reindeer herd on Taimyr Peninsula', in: *Siberian Times*, 7 February 2017, available at: http://siberiantimes.com/ecology/casestudy/features/f0285-how-poaching-is-killing-off-the-worlds-largest-reindeer-herd-on-taimyr-peninsula/.

[29]Cf. Riseth/Tømmervik/Bjerke, '175 years of adaptation' 2016.

[30]Mustonen/Jones, *Reindeer Herding* 2015, 3.

[31]A, somewhat, similiar approach to herding is also found among some shepherding communities in Britain's Lake District, see Rebanks, *Shepherd's Life* 2016, 9 et seq.

The situation in Sweden illustrates rather dramatically the need for national law-makers to respect traditional cultures as all indigenous land rights are dependent on reindeer herding, Swedish law not recognizing other forms of indigenous land rights, e.g. for fishing.

All over the European Arctic, reindeer herding cannot be understood completely without looking at the problems and challenges surrounding indigenous land rights,[32] be it full ownership or usage rights.[33] Usage rights are particularly relevant in areas where the state claims ownership of large areas of land, like in Norway and Finland.[34] Ancient indigenous ways of life, which are based on many generations of land use, often conflict with modern, state-centered, approaches to land use and land use conflicts.[35] Often it will be difficult, if not impossible, for Sámi reindeer herders to provide evidence which holds up in today's courts to show this long history of land use of non-owned lands for herding purposes.[36]

In Finland, the right to herd reindeer is not even restricted to indigenous persons or groups but in theory every EU citizen can join a reindeer herding association. The traditional Sámi methods of reindeer herding are being replaced by methods favored by the Association of Reindeer Herding Districts and the Ministry of Agriculture,[37] both of which are dominated by non-indigenous actors. This difference in approaches to reindeer herding has consequences for animals, humans and the wider environment. While the modern, non-indigenous, approach places emphasis on food production,[38] "Sámi reindeer herders [...] see reindeer herding usually in a more holistic sense, as a way of life in which the economic, ecological, social, and cultural aspects are closely intertwined. Reindeer herding is seen as an inseparable part of life"[39] and Sámi culture.

Other concerns are even more pressing from the perspective of indigenous reindeer herders: while climate change is a long-term problem, which already has direct effects on reindeer herding[40] and which devalues traditional indigenous knowledge,[41] the more immediate threat to both animals and indigenous reindeer herders stems from increased land use conflicts. Mining, forestry and the construction of infrastructure not only take away the possibility for reindeer to roam freely, they also endanger the natural environment and reduce food security for the animals and therefore also for the Sámi reindeer herders. Animal welfare is scarcely taken

---

[32]Heikkilä, *Reindeer Talk* 2006, 96.

[33]Cf. Heikkilä, *Reindeer Talk* 2006, 97.

[34]Ibid., 97.

[35]Ibid., 98 et seq.

[36]See for example European Court of Human Rights, *Handölsdalen Sami Village and others v. Sweden*, Application No. 39014/04, Judgment of 30 March 2010.

[37]Heikkilä, *Reindeer Talk* 2006, 59.

[38]Ibid., 90.

[39]Ibid., 93.

[40]See Tyler et al., 'Saami reindeer pastoralism 2007, 194 et seq.

[41]Ford/Smit/Wandel, 'Vulnerability' 2006, 150.

into account, which might also be due to a lack of information among decision-makers in Stockholm, Oslo and Helsinki despite the wealth of already existing scientific research on reindeer. This situation reflects the relative economic importance of reindeer herding: from a national perspective, reindeer herding is a "marginal industry"[42] and indeed in Finland (as opposed to Norway[43]) it is not recognized as a "formal"[44] form of income but "it appears to be mostly a curiosity".[45] On the local level, however, in particular in the border region between Finland and Norway, reindeer herding remains an important source of income and field of employment.[46] The consideration given by the authorities to reindeer herders, especially in Norway, where only Sámi herd reindeer, has had the detrimental side effect that interventionism has made reindeer herding less sustainable than it could be: "The Norwegian political system has generally failed to consider reindeer herding as a business. Rather the preferred solution of successive governments of all political colours to the problems facing the herding industry has been to increase the level of economic support to the Saami herders, in effect putting them on the dole."[47] The loss of accessible land cannot be compensated easily and so far measures taken by states have not had the effect of (re-)creating an environment in which sustainable reindeer herding is possible.

5 Concluding Remarks

Animals play important cultural, economic and spiritual roles for indigenous communities. However, contemporary law does go far enough in recognizing this particular form of interaction between humans and animals. Treating indigenous livelihoods like farming or ranching like non-indigenous counterparts means that significant cultural aspects are overlooked and that indigenous knowledge regarding the wellbeing of animals is often disregarded by political decision-makers. Climate change and increased land uses already threaten traditional ways of life in the North. The situation is worsened by the continued separation of the Sámi people by state borders. Today, there is no single form of Sámi reindeer herding but instead there are stark differences between the four states. The same applies to a great extent to fishing. The borders between the countries ruling Sámi (with the exception of Russia) were opened in the twentieth century. Yet, similar to the situation in North America despite the 1794 Jay Treaty, the borders are still very real for indigenous peoples. The way of life of many indigenous communities is dependent on the

---

[42]Heikkilä, *Reindeer Talk* 2006, 90.

[43]Ibid., 90.

[44]Ibid., 90.

[45]Ibid., 90.

[46]Ibid., 91.

[47]Reinert, *The Economics* 2006, 537.

behavior of animals. Migratory animals are not constrained by international borders. In many places around the world, pastoralist (semi-)nomadic herding has conflicted with sedentary agriculture for many centuries. Today, conflicts between settlers and indigenous communities are usually understood from the perspective of the dominant part of society, that is, the settler community (many members of which are unaware of the historical background and the land rights of indigenous peoples). In many cases, such conflicts turn on the needs of different animals, for example free-ranging reindeer as opposed to grazing cattle or other animals usually associated with farming. Through the dominance of the settler society, the needs of farming- and ranching-related animals, as well as other interests such as the construction of infrastructure and other economic activities that require land, are given precedence over the interests of wild or semi-domesticated animals on which the livelihoods of indigenous people depend. Often specific indigenous needs are seen as conflicting with the interest of the state, which means that politically, indigenous interests can be framed as conflicting with what is perceived as the 'common' good.

This risk is particularly acute in the Nordic countries due to a political and legal emphasis on equality, which leads to widespread political opposition to what is perceived as additional rights for special interest groups. While this is changing, the move towards better protection of indigenous rights is not a straightforward process. Discrimination against Sámi persons is still widespread, despite the general commitment to human rights in Norway, Finland and Sweden. A limited recognition of indigenous rights also has direct effects on the well-being of animals associated with indigenous communities.

References

Åsbakk, K., Kumpula, J., Oksanen, A., & Laaksonen, S. (2014). Infestation by *Hypoderma tarandi* in reindeer calves from northern Finland - Prevalence and risk factors. *Veterinary Parasitology, 200*, 172–178.

Brännlund, I., & Axelsson, P. (2011). Reindeer management during the colonization of Sami lands: A long-term perspective of vulnerability and adaptation strategies. *Global Environmental Change, 21*, 1095–1105.

Broadbent, N. D. (2010). *Lapps and Labyrinths: Saami prehistory, colonization and cultural resistance*. Washington, DC: Smithsonian Institution Scholarly Press, ebook, position 4137.

Burkhard, B., & Müller, F. (2008). Indicating human-environmental system properties: Case study northern Fenno-Scandinavian reindeer herding. *Ecological Indicators, 8*, 828–840.

Elbo, J. G. (1954). The Norwegian-Finnish reindeer fence. *Polar Record, 7*, 73–74. Available at: https://doi.org/10.1017/S0032247400043138.

Ford, J. D., Smit, B., & Wandel, J. (2006). Vulnerability to climate change in the Arctic: A case study from Arctic Bay, Canada. *Global Environmental Change, 16*, 145–160.

Fuglerud, Ø. (2016). Positioned creativity: Museums, politics and indigenous art in British Columbia and Norway. In M. Svašek & B. Meyer (Eds.), *Creativity in transition - Politics and aesthetics of cultural production across the globe* (pp. 158–184). New York: Berghahn.

Heikkilä, L. (2006). *Reindeer talk: Sámi reindeer herding and nature management*. Rovaniemi: Lapland University Press.

Kent, N. (2014). *The Sámi peoples of the North: A social and cultural history.* London: Hurst & Company.

Koivurova, T. (2011). Jurisprudence of the European Court of Human Rights regarding Indigenous peoples: Retrospects and prospects. *International Journal of Minority and Group Rights, 18,* 1–37.

Lantto, P. (2010). Borders, citizenship and change: The case of the Sami people, 1751–2008. *Citizenship Studies, 5,* 543–556.

Lehtola, V.-P. (2010). *The Sámi People: Traditions in transition.* Aanar/Inari: Kustannus-Puntsi.

Mustonen, T., & Jones, G. (2015). *Reindeer Herding in Finland.* Lampeter: European Forum on Nature Conservation and Pastoralism. Available at: http://www.snowchange.org/pages/wp-content/uploads/2015/06/Reindeer-herding-in-Finland.pdf

Rebanks, J. (2016). *The Shepherd's Life - A tale of the Lake District.* London: Penguins Books.

Reinert, E. S. (2006). The economics of reindeer herding - Saami entrepreneurship between cyclical sustainability and the powers of state and oligopolies. *British Food Journal, 108,* 522–540.

Riseth, J. Å., Tømmervik, H., & Bjerke, J. W. (2016). 175 years of adaptation: North Scandinavian Sámi reindeer herding between government policies and winter climate variability (1835–2010). *Journal of Forest Economics, 24,* 186–204.

Roturier, S., & Roué, M. (2009). Of Forest, Snow and Lichen: Sámi Reindeer Herders' Knowledge of Winter Pastures in Northern Sweden. *Forest Ecology and Management, 158,* 1960–1967.

Seurujärvi-Kari, I., & Carpelan, C. (2005). Regions and borders. In U.-M. Kulonen, I. Seurujärvi-Kari, & R. Pulkkinen (Eds.), *The Saami: A Cultural Encyclopedia* (pp. 294–295). Helsinki: Suomalaisen Kirjallisuuden Seura.

Snatic, J. *Hunting and Gathering by the Sami.* Available at: https://www.laits.utexas.edu/sami/diehtu/siida/hunting/jonsa.htm.

Stoyanova, I. L. (2013). The Saami facing the impacts of global climate change. In R. S. Abate (Ed.), *Climate change and Indigenous Peoples: The search for local remedies* (pp. 287–312). Cheltenham/Northampton: Edward Elgar.

Tyler, N. J. C., et al. (2007). Saami reindeer pastoralism under climate change: Applying a generalized framework for vulnerability studies to a sub-arctic social-ecological system. *Global Environmental Change, 17,* 191–206.

Warg Næss, M., Bårdsen, B.-J., Fauchald, P., & Tveraa, T. (2010). Cooperative pastoral production — the importance of kinship. *Evolution and Human Behavior, 31,* 246–258, 249.

Stefan Kirchner is Associate Professor of Arctic Law at the Arctic Centre of the University of Lapland in Rovaniemi, Finland, a visiting senior researcher at the Faculty of Law of Vytautas Magnus University in Kaunas, Lithuania, and invited professor (Jean Monnet Module) at V. N. Karazin University in Kharkiv, Ukraine. He has been visiting professor for Transitional Justice at the Faculty of Law of the University of Torino, Italy. Formerly a practicing lawyer in Germany, his current research interests include the governance of marine spaces in the Arctic and areas beyond national jurisdiction, in particular space law.

Chapter 7
China's Legal Response to Trafficking in Wild Animals: The Relationship between International Treaties and Chinese Law

Jiwen Chang

Abstract The chapter gives an account of China's new legal framework (particularly the Wild Animal Protection Law of 2016). This comprises novel official decrees which interpret the criminal law, law enforcement activities (partly police operations conducted jointly with other states), new injunctions banning ivory products, and finally criminal prosecutions. However, gaps and deficiencies persist in China's law on the books, enforcement remains slow and patchy, and international cooperation is not strong. Chang suggests several concrete measures for improvement, including the introduction of public interest litigation, better coordination among governmental departments, a trading information platform, and consultation with the secretariat of the Convention on International Trade in Endangered Species of Wild Flora and Fauna (CITES), in order to bring the Chinese legal and administrative framework fully in line with CITES.

1 Introduction

In China, the wild animals and animal products that are sold through illegal trafficking are mainly those that can be made into medicines; are raw materials in the form of ivory, rhinoceros horns, and turtle shells; and are edible or have ornamental value, such as birds, monkeys, turtles, and lizards. Due to its rapid economic development over the past decade, China has become one of the world's largest wildlife markets. The main reasons for trafficking are a lack of viable

Revised version of the original published article "China's Legal Response to Trafficking in Wild Animals: The Relationship between International Treaties and Chinese Law" by Jiwen Chang, American Journal of International Law Unbound, Volume 111, 2017, pp. 408–412. The original article was published as an Open Access article, distributed under the terms of the Creative Commons Attribution licence (http://creativecommons.org/licenses/by/4.0/).

J. Chang (✉)
Chinese Academy of Social Sciences, Institute of Law, Beijing, China

© The Author(s) 2020
A. Peters (ed.), *Studies in Global Animal Law*, Beiträge zum ausländischen öffentlichen Recht und Völkerrecht 290,
https://doi.org/10.1007/978-3-662-60756-5_7

71

substitutes for raw materials used in traditional Chinese medicines (e.g., bear bile, bear bile powder, pangolin, and other products); a preference in traditional food culture for delicacies made from wildlife; and of the private consumption by some rich and corrupt government officials of tiger's meat, bear's paw, pangolin and other wild animal products—bear's paw and pangolin being the most popular. This type of wild animal trafficking endangers the safety of animal species protected by the Convention on International Trade in Endangered Species of Wild Fauna and Flora (CITES), and damages the international image of the government and people of China. Since 2013, under the frame of construction of ecological civilization, China has taken stricter measures on legislation, administrative enforcement, judicial adjudication, and international cooperation on prevention and punishment of illegal trafficking.

2   China's Law Enforcement and Judicial Measures Against the Trafficking of Wild Animals and Their Products

2.1   Law Enforcement Measures

China joined CITES on December 25, 1980, and on April 8, 1981, CITES came into force for China. China has increased its efforts to implement CITES with the enactment of the Wild Animal Protection Law and the creation of special customs and criminal regulations.

The Chinese government has strictly implemented wild animal protection laws and regulations, and now cracks down on trafficking, defends the honour of the nation, and protects the ecological balance. From January 6 to February 5, 2013, China and 21 other Asian and African countries organized an operation, codenamed "Cobra," against international wildlife crime. The operation resolved 71 cases, arrested 85 suspects, and seized 185 kg of ivory, 13 kg of rhino horns, about 50 kg of pangolin scales, and 76 rhinoplax vigil beaks. Operation "Cobra II," coorganized with twenty-eight Asian, African, and North American countries, was conducted from December 30, 2013 to January 26, 2014. It resulted in the seizure of 286 kg of ivory, 802 pieces of reticulated python skin, and 120 kg of pangolin scales.[1] In January 2014, the Chinese government publicly destroyed 6.1 tons of confiscated ivory in Dongguan City of Guangdong province for the first time, and in the year after, the Chinese government, again publicly, destroyed 662 kg of ivory.[2]

---

[1]Chinese State Forestry Administration, 'China Customs' Special Action on Combating Wildlife Smuggling', 9 June 2015, available at: http://www.forestry.gov.cn/wlmq/3585/content-773841.html.

[2]Yang, 'Save Endangered Wildlife, Fight Against Trafficking, We Are on the Way' 2015.

2.2   Judicial Measures

China's consciousness of ecological civilization and laws will need to improve over time. Although the Chinese legislature has banned ivory imports with the Wild Animal Protection Law, wildlife poaching and trafficking have yet to be eliminated entirely. Several typical criminal cases in recent years have, however, had a significant social impact. In November 2014, a farmer in Henan Province was sentenced to three months detention for capturing 87 toads. The farmer was the first person to be punished in such a case.[3] Two years later, some farmers were also sanctioned for capturing frogs. In November 2015, a college student from Henan province was sentenced to 10.5 years in prison for hunting and selling hobby falcons. These activities had not been previously treated as crimes. With constant public education and strict law enforcement, wildlife poaching and trafficking have been greatly reduced. Young people in particular tend to have lost interest in obtaining wild animals and their products. In the long run, poaching and trafficking will therefore likely continue to decline.[4] In China's vast rural areas, wild animals that had largely disappeared for two or three decades have begun to reappear, indicating the effectiveness of China's ecological protections.

3   New Chinese Legislation on the Elimination of Trafficking of Wild Animals and Their Products

3.1   Interpretation of Crimes Against Wild Animals by the Standing Committee of the National People's Congress

In 1997, China's criminal law stipulated conditions and penalties only for illegal hunting and killing, acquiring, transporting, selling, and trafficking precious and endangered wild animals and their products that are significantly protected by the state. As consumption drives these crimes, it is also necessary to combat illegal purchases in order to effectively protect wild animals from illegal hunting, transport, trafficking, and other such acts. On April 24, 2014, the Standing Committee of the National People's Congress adopted "Interpretation of Article 341, Article 312 under Criminal Law of the People's Republic of China," which specifies that it is a crime to purchase as food wildlife and wildlife products that are under special state protection. On February 6, 2017, China's internet spread a Hong Kong businessman's blog that showed that local government officials in Guangxi had

---

[3]Li, 'Capturing 87 Toads, A Farmer Gets Criminal Detention for Three Months' 2014.

[4]With the development of the internet and we-media such as we-chat, more and more cases have been reported to the government or exposed to the public. In addition, more and more cases are treated as criminal. This does not mean, however, that the occurrence of crimes such as poaching and trafficking is actually increasing.

invited him to eat pangolin dishes in their office. The blog entry was shared quickly and led to condemnation. In February 2017, the businessman and the chefs who purchased pangolin were criminally detained on suspicion of illegal acquisition of precious, endangered wildlife.

3.2   The Wild Animal Protection Law and Its New Provisions

In order to combat the trafficking and illegal trade in wild animals at the source as well as the intermediate steps, the Wild Animals Protection Law (2016) prohibits the following activities: (a) producing or selling food that uses wild animals under national special protection and the products of such animals; (b) producing or selling food using wild animals not specially protected by the state and the products of such animals without evidence of the legal source; (c) illegally purchasing for food wild animals under national special protection and the products of such animals; and (d) illegally advertising wild animals and their products.

As regards international cooperation in the fight against wildlife trafficking, Article 36 of the Wild Animals Protection Law provides that the state has to organize international cooperation and communication activities regarding wildlife protection and related enforcement activities, to establish coordination mechanisms to prevent and combat trafficking and illegal trade in wild animals and their products, and to conduct operations to combat trafficking and illegal trade. This provision lays down the legal foundation for China to better fulfil its duties under CITES and to combat wildlife smuggling more comprehensively.

As regards public participation, Article 5 of the Wild Animals Protection Law stipulates that "the State encourages citizens, individuals and other organizations to support wildlife protection public welfare activities by participating in wildlife protection activities through donation, subsidy and volunteer service." At present, "Let Birds Fly," "China Finless Porpoise Protection Action Network," "Friends of Nature," and other NGOs as well as some environmental activists, participate in the protection of birds. Their work attracted Chinese Central Television, the local television, and other news media to support, or to join in, thereby helping the public security bodies to arrest suspects. For example, on February 26, 2017, the Guangxi police seized seventy wild turtles in a rented house[5] on the border between China and Vietnam based on a tip.

3.3   Ban on Ivory and Its Products in Recent Years

Since the end of 2012, the Chinese government has advocated ecological civilization and has strengthened its international cooperation on wildlife protection. On February 26, 2015, the State Forestry Administration announced that China had

---

[5]Yan/Huabin/Qiuhong, 'Suspected of Smuggling Wildlife' 2017.

implemented two temporary injunctions: first, a prohibition on African ivory carvings after CITES entered into force; second, a prohibition on the import of memorial ivory originating from hunting in Africa. During the import injunction, the State Forestry Administration had to suspend the relevant administrative licensing matters. On March 2, 2015, Chinese President Xi Jinping met Prince William of Great Britain and introduced him to China's policy and to its work on protecting wild animals such as elephants, hoping to enhance international cooperation in this field.[6] Two days later, Prince William visited Xishuangbanna, a city in the Yunnan province, to examine the status of elephant protection.[7] On September 25, 2015, Xi Jinping and United States President Barack Obama, as leaders of the two major destination countries for ivory trading, agreed to enact a domestic ban. Each country is required to take effective and timely action to ensure the gradual cessation of ivory import and export trade (except in a few special cases). On December 2, 2015, Xi Jinping visited Zimbabwe's wild animal rescue base. Xi Jinping committed China to continue to support the strengthening of the protection of wild animals through material assistance, the exchange of experience and so on. It is evident from these undertakings that China's diplomacy and international cooperation on wildlife protection has changed from passive adaptation to active participation.

When the two temporary injunctions on ivory products expired in March 2016, the State Forestry Administration immediately announced more stringent control measures for imports of ivory and its products. The implementation of the two temporary injunctions has been extended to December 31, 2019, and the scope of prohibited imports of ivory and its products has been expanded.

Where there is a market, there will inevitably be some overseas trafficking and poaching. The General Office of the State Council promulgated the Notice of Stopping Commercial Production and Sales of Ivory and Products by decree on December 30, 2016, requiring the closure of commercial processing and sales of ivory and its products in stages and in groups. Specifically, China shut down several processing and sales units for ivory and its products by March 31, 2017 and will fully stop the relevant activities by December 31, 2021. This measure will put many ivory engravers out of work, and many lawful collectors will have difficulty selling the ivory art works that they own, leading to huge losses. There has been to date no information on government compensation for ivory engravers. However, no ivory engraving companies and operating companies are heard to complain in public as of July, 2018. It is thus clear that Chinese society has a consensus on wildlife conservation. Although Donald Trump, the President of the United States, decided to relax restrictions on ivory imports from some African countries, The President Xi of China acted as a firm environmentalist. The Chinese government has no signs of lifting the ban of the international and domestic ivory trade as of July, 2018. Some international organizations commented that the Chinese government is becoming one of the world

---

[6]Liu, 'Xi Jinping Met with Prince William' 2015.
[7]Hu/Yang, 'British Prince William Visits Xishuangbanna' 2015.

leaders in the field of wildlife conservation.[8] Thus, in order to protect the African elephants, China has paid a huge price.

4   Suggested Improvements to China's Legal Response to Trafficking

First, the Chinese wildlife protection list is not consistent with the CITES appendix. The Wild Animal Protection Law protects wildlife by classification, focusing on precious, endangered wild animals. National specially protected wild animals are divided into two classes of wildlife protection. The list of national specially protected wild animals must be formulated and adjusted every 5 years, based on the scientific assessment of the department of wildlife protection as approved by the State Council. Some wild animals highly valued in the market, such as black bears and pangolins, are among the protected animals listed in Appendix I of CITES, but only classified as wildlife under the second class protection under Chinese law, making their level of protection lower than it should be. As pangolins are endangered in China, these should be included in the first class protection list, as should wild black bears. These adjustments would help to crack down on trafficking.

Second, protection levels sometimes differ between wildlife and artificially bred wild animals. A very controversial provision in Article 28 of the Wild Animals Protection Law requires wild animals with mature and stable artificial breeding technology to be entered on the list of protected artificially bred wild animals formulated by the State Council wildlife protection authorities based on scientific justification. According to the relevant wild population protection cases, the artificial population of such wild animals may no longer be included in the national specially protected wildlife list and is subject to different management measures than the wild population. According to Article 28, if the black bear's artificial breeding technology has matured and the number of wild black bears is relatively stable, then artificially bred black bears are not to be treated as wild animals, and thus illegally purchasing or trafficking artificially bred black bears will not constitute the illegal acquisition or trafficking of precious and endangered wildlife. This reasoning may diverge from the intent of CITES. CITES provides that reared tigers, bears, and other animals continue to be included in the national specially protected wildlife list. Moreover, research into artificial substitutes for bear bile should be strengthened. China should prohibit the extraction of bear bile within fifteen years and gradually eliminate trafficking in black bear and bear bile products.

Third, a lack of regulatory personnel and a limited regulatory capacity makes it difficult to find and combat all trafficking. One suggestion is that wildlife protection regulators and public security organs promote public participation and supervision by establishing measures to reward reporting. The information that the public submits

---

[8]Bo/Ke, 'Chinese government is becoming one of the world leaders in the field of wildlife conservation' 2018.

must be taken seriously or the relevant manager should be held accountable. Other suggestions are establishing a shared platform for regulatory information across various departments and regions and directing the interdepartmental and interregional regulatory cooperation mechanism toward joint or collaborative law enforcement. Furthermore, in the next reform of the Wildlife Protection Law, clear provisions for a system of public interest civil litigation should be proposed. This would allow social organizations and individuals to prosecute the trafficking, poaching, sales, and transport of wild animals and their products and to make offenders liable for ecological damage.

Fourth, some law enforcement agencies are inactive or slow to act when dealing with illegal activities. Some offenders even sell wild animals and their products near local industrial and commercial departments, public security police stations, and county or town governments without being subject to any control. To solve the prevarication between the local government and the regulatory authorities, the state should first establish a list of regulatory powers for the local government and for the departments of forestry, public security, customs, business, and environmental protection. The state would provide that these no longer be exempt from liability for dereliction of duty in accordance with this list. China should establish a system for assessing local government's performance annually with respect to wild animal protection. Local governments should be held accountable for abusive action, slow action, and inaction by law enforcement agencies. A further suggestion is that in the next amendment of the Wild Animal Protection Law, clear provisions for a system of public interest administrative litigation be proposed. This would allow social organizations and individuals to sue local government and its regulatory authorities for abusive action, slow action, and inaction. Only in this way can these authorities be compelled to supervise in keeping with the law. In addition, a third-party assessment system is needed. NGOs and other third parties could thereby assess the government's performance as regards law enforcement.

Fifth, the mechanism for international antitrafficking cooperation needs to be refined. Under the coordination of the State Forestry Administration, NGOs at home and abroad should strengthen their communication and cooperation in the protection of wild animals and the fight against trafficking, poaching, illegal transport, and sales by forming an information network to combat trafficking of wild animals beyond national borders. As coordinated by the State Council, the departments of forestry, industry and commerce, customs, environmental protection, and public security should jointly create a list of regulatory powers in the fight against the smuggling and illegal trade of wild animals and their products. This would refine the coordination mechanism among departments in order to establish a unified information platform regarding wildlife in international and domestic trade. Furthermore, under the leadership of the Ministry of Foreign Affairs, a mechanism for international negotiations and cooperation and a unified information platform regarding transnational law enforcement should be collaboratively established, based on the coordination of the departments of forestry, industry and commerce, customs, environmental protection and public security. In addition, the Chinese government should consult with the CITES Secretariat and in particular should strengthen communication with the countries of origin and transport of wild animals and their products. As regards cross-border trafficking, China's communication, coordination, and notification mechanism should be improved jointly by signing bilateral or multilateral treaties with important neighbouring countries and major trading partners and by establishing the mutual convergence of law enforcement. Only in this way can trafficking and illegal trade be systematically curbed.

Finally, the government should plan to compensate ivory engravers for ceasing the commercial production and sales of ivory and products and help them make the transition to other livelihoods. Without this assistance, black markets will continue to exist, at least in the near future.

5 Conclusion

China now has strict legislation on the protection of wild animals. The next step is to enforce the law through the measures discussed. China should strengthen its publicity, education, and social participation on these issues and ensure that law enforcement is continuous, strict, and uniform across all districts. China's international cooperation in wildlife protection would thereby achieve better results. Moreover, it is only in this way that China can gain international recognition for acting as an environmentally responsible country.

References

Bo, W., & Ke, T. (2018, April 28). World Animal Protection, 'Chinese government is becoming one of the world leaders in the field of wildlife conservation'. *The Paper.*

Hu, C., & Yang, Y. (2015, March 4). British Prince William Visits Xishuangbanna and has "Intimate Contact" with Elephants. *Xinhua.*

Li, Z. (2014, December 1). Capturing 87 toads, a farmer gets criminal detention for three months. *Dahe.*

Liu, H. (2015, March 3). Xi Jinping Met with Prince William. *Xinhua Daily Telegraph.*

Yan, C., Huabin, Z., & Qiuhong, H. (2017, March 1). Suspected of smuggling wildlife, a Man's Bribe of 300 Thousand Yuan seeking "Exception" was rejected'. *Guanxi News.*

Yang, S. (2015, July 3). Save endangered wildlife, fight against trafficking, we are on the way. *China Daily.*

Jiwen Chang is a deputy director at the Research Institute of Resources and Environment Policies of Development Research Center of the State Council as well as a professor at the Institute of Law of Chinese Academy of Social Sciences. He is also a part-time professor at China University of Geosciences (Beijing) and Capital University of Economics and Business. He is a member of the Standing Committee of Beijing Municipal Peoples' Congress. His current research interests relate to animal welfare law, environmental policy and law.

Chapter 8
Corruption Gone Wild: Transnational Criminal Law and the International Trade in Endangered Species

Radha Ivory

Abstract The chapter sketches how the topics of corruption and endangered animal trafficking have been intertwined in hard and soft international law, including by United Nations Security Council resolutions. The legal documents depict corruption as enabling the illegal wildlife trade, and, concomitantly, portray the illegal wildlife trade as prompting official corruption. Ivory cautions against linking the two legal frameworks and reform agendas. Notably, the linkage implies that animal products are legitimate commodities when traded in an uncorrupted global market. The linkage may also focus too much attention on the criminal individuals who contribute to animal extinction, rather than on the large-scale environmental changes caused by industrialization and urbanization. Finally, the twinning of the two discourses could amplify the demonization of low-level bribery and poaching that are typically associated with the Global South. A combined anti-corruption/wildlife trafficking debate may distract from the opportunities for illicit investment and excessive consumption in the Global North, which enable and drive the crimes.

1 Introduction

If violence is an obvious *modus operandi* of the illegal wildlife trade, corruption is an increasingly visible other. At least since the mid-2000s, states have softly acknowledged connections between the transnational crimes and called on each

---

Revised version of the original published article "Corruption Gone Wild: Transnational Criminal Law and the International Trade in Endangered Species" by Radha Ivory, *American Journal of International Law Unbound*, Volume 111, 2017, pp. 413–418. The original article was published as an Open Access article, distributed under the terms of the Creative Commons Attribution licence (http://creativecommons.org/licenses/by/4.0/). The revisions were finalised on 1 June 2018.

R. Ivory (✉)
University of Queensland, TC Beirne School of Law, St Lucia, QLD, Australia
e-mail: r.ivory@law.uq.edu.au

© The Author(s) 2020
A. Peters (ed.), *Studies in Global Animal Law*, Beiträge zum ausländischen öffentlichen Recht und Völkerrecht 290,
https://doi.org/10.1007/978-3-662-60756-5_8

other to uphold conventions that support the respective fields of cooperation. In recent years, the Security Council has hardened the doctrinal links in resolutions on particular 'security' situations. These constructions, in turn, resonate with research publications by academics, international organisations, and non-state actors. Examining this emerging orthodoxy, this chapter briefly sketches how, and with what effect, the problems of corruption and endangered animal trafficking have been linked in international law. It first compares and contrasts the 'hard law' legal frameworks on corruption and on animal trafficking. After that, it illustrates how those two regimes have been related in international reports, non-binding international instruments, and UN Security Council Resolutions. Finally, drawing on critical literatures in international law and criminology, it cautions against an automatic merger of these agendas for global law enforcement and reform. Like other transnational criminal laws, the anti-corruption treaties have practical limitations, ideological biases, and potentials for unintended consequences. These features qualify their utility as 'tools' in the fight for animal welfare. They may also mask the ways in which efforts to prevent and suppress wildlife trafficking are both anthropocentric and sources of human insecurity.

2 Prohibiting Corruption and Wildlife Trafficking Through International Law

Illegal wildlife trafficking and corruption are dealt with under different international agreements with different logics of regulation. The key global *Convention on International Trade in Endangered Species of Wild Fauna and Flora* (CITES)[1] aims to regulate the transactions in endangered animals and plants by establishing an international licensing system.[2] The CITES appends three lists of variously at-risk species. Before listed animals or their parts or products ('specimens') can be lawfully traded, states must demand certain documentation.[3] Those documents may only be issued when approvals have been obtained from national 'Scientific' or 'Management Authorities'.[4] States have discretion as to how they enforce the CITES system within domestic law; however, they must penalise the illegal trade and/or possession of listed specimens and enable the confiscation or return of such specimens to export countries.[5] By contrast, the international regime against corruption contains more

---

[1]Convention on International Trade in Endangered Species of Wild Fauna and Flora (CITES), 3 March 1973, 993 UNTS 243.

[2]See, generally, Bowman/ Davies/Redgwell, *Lyster's International Wildlife Law* 2010, 14, ch. 15.

[3]CITES 1973, arts. III–IV.

[4]Ibid. and art. IX.

[5]CITES 1973, art. VIII(1).

than a dozen binding instruments addressed to offences like bribery, embezzlement, and money laundering.[6] These anti-corruption treaties commit their state parties to criminalising abuses of power or trust for private gain, along with activities that enable offenders to avoid prosecution and/or enjoy ill-gotten wealth.[7] States must meet these transgressions with criminal penalties and take steps to ensure that persons can be deprived of illicit wealth, as a rule. Assuming that assets, offenders, and evidence may be located in different states, the treaties commit state parties to assisting each other in criminal matters.[8] In this way, the anti-corruption treaties are examples of 'suppression conventions', which are said to constitute the hard core of a 'Transnational Criminal Law'.[9]

3 Linking International Norms Against Corruption and Wildlife Trafficking

The treaties on wildlife trafficking and corruption thus use different approaches to controlling acquisitive crimes of globalisation. Textually they are connected through oblique references to sustainable (economic) development in some anti-corruption standards.[10] Links between wildlife trafficking and corruption are drawn rather in research publications, soft laws, and Security Council resolutions, which I discuss below. Taken together, these documents paint a picture of corruption as enabling the illegal wildlife trade and the illegal wildlife trade as prompting official corruption. Both types of crime are tied to inequality between countries and instability within them.

3.1 Research Publications

Academics, international and non-governmental organisations have presented corruption and the trafficking of wild animal and plant specimens as functionally

---

[6]See, esp., Convention on Combating Bribery of Foreign Public Officials in International Business Transactions, 17 December 1997, 37 ILM (1998), 1 (OECD Convention); United Nations Convention against Transnational Organized Crime, 15 November 2000, 2225 UNTS 209 (UNTOC); United Nations Conventions against Corruption, 31 October 2003, 2349 UNTS 41 (UNCAC). See, further, Ivory, *Human Rights of Bad Guys* 2014, 1–3, 16 (table 2.1).

[7]Ivory, *Human Rights of Bad Guys* 2014, 12–22.

[8]See, generally, Ivory, *Human Rights of Bad Guys* 2014, ch. 4.

[9]See, esp., Boister, '"Transnational Criminal Law"?' 2003, 953–976.

[10]See, esp., UNCAC 2003, preamble; UNTOC 2000, art. 30(1); OECD, Recommendation of the Council for Further Combating Bribery of Foreign Public Officials in International Business Transactions, C(2009)159/REV1/FINAL, 26 November 2009, preamble. See also UN General Assembly, Transforming Our World: The 2030 Agenda for Sustainable Development, 25 September 2015, A/RES/70/1, Annex, Goals 15.7 and 16.5.

interconnected.[11] The relationship is complex, not least at the micro-level of particular crime situations.[12] Generalising here about the macro-level, the CITES emerges
as a tool for animal protection but also a source of opportunities and incentives for
bribery.[13] On the one hand, wildlife control regimes give public officials monopoly
rights to authorise or otherwise enable lucrative transactions.[14] Officials may have
broad (and/or weakly supervised) discretions to deploy their powers.[15] Other (more
senior) persons may resist implementing the relevant laws or making reforms that
would help combat corruption or improve the operation of protections for animals.[16]
They may profit as illicit 'producers' or 'traders' in their own right or as beneficiaries
of bribes paid or raised by others in public service.[17] On the other hand, traffickers
have economic reasons to offer illegal premiums to authorities or to seek to have
permission processes expedited, conditions eased, and/or paperwork falsified. Alter-
natively, officials may be bribed to ignore illegal operations and/or to warn traffickers about planned interception efforts or raids.[18] Some traffickers could also
operate with, or as, organised criminal groups.[19] Finally, corruption and wildlife
trafficking are seen to enable, and be enabled by, the same global disparities in

---

[11]Wyatt/Cao, 'Corruption and Wildlife' 2015, 7, available at: www.u4.no/publications/corruption-
and-wildlife-trafficking/downloadasset/3832. See, e.g., Environmental Investigation Agency (EIA),
'Time for Action: End the Criminality and Corruption Fuelling Wildlife Crime' (2016), available at:
www.eia-international.org; UN Office on Drugs and Crime (UNODC), 'World Wildlife Crime
Report: Trafficking in Protected Species' (2016), available at: https://www.unodc.org/unodc/en/
data-and-analysis/wildlife.html; Moreto/Brunson/Braga, 'Law Enforcement Ranger Wrongdoing'
2015, 359–380; Smith/Walpole, 'Should Conservationists Pay More Attention to Corruption?'
2005, 251–256; World Wildlife Fund (WWF)/TRAFFIC Wildlife Crime Initiative, 'Strategies for
Fighting Corruption in Wildlife Conservation: A Primer' (2015), available at: wwf.panda.org/?
257350/Strategies-for-fighting-corruption-in-wildlife-conservation; van Uhm/Moreto, 'Corruption
within the Wildlife Trade' 2017; Wyatt, *Wildlife Trafficking* 2013, 52–53.

[12]See, e.g., Moreto/Brunson/Braga, 'Law Enforcement Ranger Wrongdoing' 2015, 367–376; van
Uhm/Moreto, 'Corruption within the Wildlife Trade' 2017, 8–16. See also UNODC, 'World
Wildlife Crime Report' 2015, 23–24.

[13]See generally UNODC, 'World Wildlife Crime Report' 2015, 19, 23–24.

[14]Ibid., 97.

[15]Moreto/Brunson/Braga, 'Law Enforcement Ranger Wrongdoing' 2015, 377; WWF/TRAFFIC,
'Strategies for Fighting Corruption' 2015, 8–13; Wyatt/Cao, 'Corruption and Wildlife' 2015, 8–14.

[16]WWF/TRAFFIC, 'Strategies for Fighting Corruption' 2015, 13–14; Wyatt, *Wildlife Trafficking*
2013; Wyatt/Cao, 'Corruption and Wildlife' 2015, 12.

[17]For an example from the illegal timber trade, albeit not with reference to the CITES, see Global
Witness, 'Cambodia's Family Trees: Illegal Logging and the Stripping of Public Assets by
Cambodia's Elite', 2007, available at: https://www.globalwitness.org/en/reports/cambodias-fam
ily-trees, 10–11.

[18]WWF/TRAFFIC, 'Strategies for Fighting Corruption' 2015, 10–15; Wyatt/Cao, 'Corruption and
Wildlife' 2015, 9–10.

[19]Ibid., 6. See also Ayling, 'What Sustains Wildlife Crime?' 2013, 68.

'development' and 'governance'.[20] Hence, addressing one problem may contribute to efforts to ameliorate the other.

3.2 Soft Law Statements

States have called on each other to address the problems of wildlife-related corruption through non-binding international decisions. At least since 2007, bodies within the UN system have called on countries to use the UN Conventions against Corruption and Organized Crime (UNCAC and UNTOC) to combat animal trafficking.[21] Moreover, in 2010, the CITES Secretariat formed an International Consortium on Combating Wildlife Crime that includes the UN Office on Drugs and Crime and the World Bank[22]: the latter organisations are intimately engaged with worldwide anti-corruption activity. Three years later, in December 2013, the General Assembly '[e] mphasis[ed] that coordinated action is critical to eliminate corruption and disrupt the illicit networks that drive and enable trafficking in wildlife, timber and timber products (. . .)'.[23] Less than 2 years after that, it resolved, without a vote, to call on Member States to 'prohibit, prevent and counter any form of corruption that facilitates illicit trafficking in wildlife and wildlife products', as well as to join and implement the UNCAC and UNTOC.[24] Similar sentiments have been expressed in decisions of the CITES' Conference of the Parties,[25] the UN's Economic and Social Council,[26] and the Environment Assembly of the UN Environment Programme,[27] as

---

[20]See generally EIA, 'Time for Action' 2016, 16; Smith/Walpole, 'More Attention to Corruption' 2005, 251–252; van Uhm/Moreto, 'Corruption within the Wildlife Trade' 2017, 2–4; Wyatt/Cao, 'Corruption and Wildlife' 2015, 7–9. Cf UNODC, 'World Wildlife Crime Report' 2015, 19.

[21]United Nations, Commission on Crime Prevention and Criminal Justice, Resolution 16/1, International Cooperation in Preventing and Combating Illicit International Trafficking in Forest Products, including Timber, Wildlife and other Forest Biological Resources, E/2007/30/Rev.1 & E/CN.15/2007/17/Rev.1, 23–27 April 2007.

[22]CITES Secretariat, 'Projects and Initiatives: The International Consortium on Combating Wildlife Crime' (website), available at: https://www.cites.org/eng/prog/iccwc.php.

[23]United Nations, General Assembly, Resolution 68/193, Strengthening the United Nations Crime Prevention and Criminal Justice Programme, in particular its Technical Cooperation Capacity, A/RES/68/193, 18 December 2013, preamble.

[24]United Nations, General Assembly, Resolution 69/314, Tackling Illicit Trafficking in Wildlife, A/RES/69/314, 30 July 2015, paras. 9–11. See also United Nations, General Assembly, Tackling Illicit Trafficking in Wildlife: Report of the Secretary-General, A/70/951, 16 June 2016, para. 56; United Nations, General Assembly, Resolution 71/326, Tackling Illicit Trafficking in Wildlife, 11 September 2017, A/RES/71/326, esp. paras. 8, 19–20.

[25]CITES, Seventeenth Meeting of the Conference of the Parties, Resolution 17.6, Prohibiting, Preventing, Detecting and Countering Corruption, which Facilitates Activities Conducted in Violation of the Convention, Conf. 17.6, 24 September–4 October 2016.

[26]United Nations, Economic and Social Council, Resolution 2013/40, Crime Prevention and Criminal Justice Responses to Illicit Trafficking in Protected Species of Wild Fauna and Flora, E/RES/2013/40, 25 July 2013.

[27]United Nations, Report of the United Nations Environment Assembly of the United Nations Environment Programme, 23–27 June 2014, A/69/25, 20–23, Resolution 1/3, para. 2(g). See also

well as at high-level meetings of international leaders.[28] For example, the G20 annexed 'High Level Principles on Combatting Corruption Related to Illegal Trade in Wildlife and Wildlife Products' to their 2017 Hamburg Declaration.[29] Later that year, the CITES' Secretary General addressed the UNCAC state parties in the opening session of their conference.[30]

3.3 Security Council Resolutions

In parallel, the UN Security Council has been reconstructing wildlife trafficking as a danger to the peace. In a significant development,[31] it has hardened the doctrinal connections between armed conflict, illegal trade in animal products, and other forms of transnational criminality. Thus, resolutions on the Central Africa Republic (CAR) and the Democratic Republic of Congo (DRC) attribute violence there, in part, to the ability of groups to fund their activities through the illegal sale of natural resources.[32] Travel bans and asset freezes are targeted at individuals and entities who, amongst other things, support the trade in elephant tusks.[33] In addition, there is an exception to an arms embargo to 'defend against poaching [and] smuggling of ivory (...)' from CAR,[34] and mention of the smuggling of Congolese ivory in the same breath as 'the importance of neutralising all armed groups' in the DRC.[35] Outside the scope of the CITES but still with relevance to animal products and biodiversity, council resolutions signal that illegal fishing and corruption are barriers

---

United Nations, United Nations Environment Assembly of the United Nations Environment Programme, Resolution 2/14, Illegal Trade in Wildlife and Wildlife Products, 23–27 May 2016, para. 1(b).

[28]London Conference on the Illegal Wildlife Trade, Declaration, 12–13 February 2014, available at: https://www.gov.uk/government/topical-events/illegal-wildlife-trade-2014, arts. IX–XII; East Asia Summit Declaration on Combating Wildlife Trafficking, 13 November 2014, available at: https://cites.org/eng/east_asia_summit, art. 14.

[29]G20, Leaders' Declaration: Shaping an Interconnected World, Annex: G20 High Level Principles on Combatting Corruption Related to Illegal Trade in Wildlife and Wildlife Products, 8 July 2017, available at: https://www.g20germany.de/Webs/G20/EN/G20/Summit_documents/summit_docu ments_node.html.

[30]United Nations, Office on Drugs and Crime, 'Press Release: Links between Corruption and Wildlife Crime Highlighted at UN Anti-Corruption Conference', 6 November 2017, available at: https://www.unodc.org/unodc/en/press/allpress.html?ref=fp.

[31]Peters, 'Novel Practice of the Security Council' 2014.

[32]See, e.g., United Nations, Security Council, Resolution 2339, S/RES/2339 (2017), 27 January 2017, preamble (on CAR); United Nations, Security Council, Resolution 2293, S/RES/2293 (2016), 23 June 2016, preamble (on DRC).

[33]See, e.g., UNSC, Resolution 2339 (2017), para. 17(e); UNSC, Resolution 2293 (2016), para. 7(g).

[34]UNSC, Resolution 2339 (2017), para. 1(f).

[35]UNSC, Resolution 2293 (2016), preamble.

to the consolidation of peace and sovereignty in Somalia[36] and Guinea-Bissau,[37] along with drug trafficking, money laundering, and piracy.

4 Contesting the Connection Between the Anti-corruption and Anti-wildlife Trafficking Agendas

Hence, there would seem to be an emerging international consensus that wildlife trafficking, transnational organised crime, and corruption must be addressed together and that their respective regimes, whilst distinct, are complementary. Yet, just as some have questioned invoking anti-corruption law to support the cause of human rights (and vice versa), so I see three reasons to pause before merging the international anti-corruption and pro-wildlife agendas wholesale. My grounds are pragmatic, normative, and epistemological.[38]

4.1 Effectiveness

First, the enforcement of international anti-corruption law is not a silver bullet for difficulties in enforcing international laws that aim to control the trade in endangered species. In fact, transnational measures against economic crime face their own critiques about effectiveness. There are initial problems with determining whether—and, if so, why—states comply with duties to implement international anti-corruption treaties. If they do so out of self-interest or peer pressure, the treaties might not be such a successful ethical check on power-political calculations.[39] Further, it is difficult to measure and prove the effectiveness of international crime control measures on individual decisions to break the law. Perhaps for this reason, inter governmental assessments of anti-money laundering controls have tended to focus on the 'volume of activities undertaken by competent authorities', even though these measures say little 'about the real impact of such actions on the criminal problem itself'.[40] Finally, even high-income countries would seem reluctant to prosecute foreign bribe payers and/or confiscate and return illicit wealth laundered

---

[36]See, e.g., United Nations, Security Council, Resolution 2125, S/RES/2125 (2013), 18 November 2013, preamble; United Nations, Security Council, Resolution 2317, S/RES/2317 (2016), 10 November 2016, preamble, paras. 17 and 21; United Nations, Security Council, Resolution 2383, S/RES/2383 (2017), 7 November 2017, preamble, para. 2; United Nations, Security Council, Resolution 2385, S/RES/2385 (2017), 14 November 2017, preamble, para. 20.

[37]See, e.g., United Nations, Security Council, Resolution 2343, S/RES/2343 (2017), 23 February 2017, preamble; United Nations, Security Council, Resolution 2404, Resolution 2404 (2018), 28 February 2018, preamble, para. 20.

[38]See, further, Ivory, 'Asset Recovery in Four Dimensions' 2017, 176–210.

[39]See Sharman, *The Despot's Guide to Wealth Management* 2017, 13–14.

[40]Vettori, 'Evaluating Anti-Money Laundering Policies 2013, 474–485. See also Halliday/Levi/Reuter (for the Centre on Law & Globalization), 'Global Surveillance of Dirty Money' 2014.

through their financial centres.[41] This may weaken any general deterrent effect of international anti-corruption measures insofar as it sends an inconsistent message to individuals (and corporations) about the likelihood that wrongdoing will be detected, investigated, and punished.

4.2 Human Rights

Second, increasing compliance with international law is not an absolute good for governance—it also generates costs. The international treaties against corruption commit states to standardising their domestic responses to certain types of behaviours and to cooperating with each other in judicial and administrative matters. Both Gless and Boister discuss the risks to defendants when states pool their *ius puniendi*.[42] Focusing on the international campaign for asset recovery, I have shown elsewhere that states' efforts to cooperate in confiscation matters may raise issues under rights to fair trials and property. International judges permit such interferences as a means of realising important law enforcement objectives; however, they insist on the conditions of lawfulness and proportionality.[43] In the wildlife protection context, human rights considerations have a communitarian dimension. Indigenous and tribal peoples have collective rights to natural resources, including when that property is contained within nature reserves.[44] In *Kaliña and Lokono Peoples v. Suriname*, the Inter-American Court of Human Rights found environmental protection and self-determination to be compatible policy objectives.[45] But the court there referred to international instruments that recognise both indigenous peoples' special interests in land and the global collective interest in biodiversity.[46] The CITES makes only general references to natural resource sovereignty and provides no traditional use exception for animal products: it is less amenable to harmonious interpretation. Hence, there is greater potential for conflict between

---

[41]Heimann/Foldes/Coles (for Transparency International), 'Exporting Corruption' 2015; Gray et al. (for the International Bank for Reconstruction and Development/World Bank/Organisation for Economic Co-operation and Development), 'Few and Far' 2014.

[42]Boister, 'Human Rights Protections in the Suppression Conventions' 2002, 199–227; Gless, 'Bird's-Eye View and Worm's Eye View: Towards a Defendant-Based Approach in Transnational Criminal Law' 2015, 117–140.

[43]Ivory, *Human Rights of Bad Guys* 2014, ch. 5 and 6.

[44]See, e.g., *Centre for Minority Rights Development (Kenya) and Minority Rights Group International on behalf of Endorois Welfare Council v. Kenya*, African Commission of Human and Peoples' Rights, 4 February 2010, Communication No. 276/03; *The Case of the Kaliña and Lokono Peoples v. Suriname*, Judgement of 25 November 2015 (Merits, Reparations, and Costs), Series C No. 309. See also *African Commission of Human and Peoples' Rights v. Kenya*, African Court on Human and People's Rights, 26 May 2017, Application No. 006/2012.

[45]*Kaliña and Lokono Peoples v. Suriname* 2015, paras. 173, 181.

[46]Ibid., paras. 174–180.

collective rights that benefit indigenous and tribal peoples and norms that protect wildlife in international law.

4.3   Politics

Third, measures to prevent animal trafficking and to control corruption may serve particular political agendas. This is not simply a point about Western states and non-state actors advocating regimes that prohibit transnational crime.[47] Rather, the anti-corruption and anti-wildlife trafficking orders may reflect, and help maintain, certain ideas about acceptable power relations between humans and animals, peoples and states. To begin, the linking of anti-corruption and wildlife trafficking discourses may potentially subjugate animals, since it suggests that animal products are legitimate commodities when traded in uncorrupted global markets. Likewise, the merger of the anti-corruption and wildlife trafficking discourses may be taken to signal that criminal individuals have caused mass animal endangerment, rather than large-scale environmental changes post-industrialisation.[48] Similarly, critical anti-corruptionists (and Third World scholars of international law) may see a postcolonial downside to a twinned agenda. The discourses of anti-corruption and anti-wildlife trafficking could each serve to demonise forms of conduct, like low-level bribery and poaching, which are perceived to be more prevalent in the Global South. They may together distract from the possibilities for illicit investment and patterns of excessive consumption in the Global North, which enable and drive the crimes.[49] Finally, some might take the Security Council's efforts to protect animals to be an example of the securitisation of global environmental governance or the militarisation of transnational organised crime controls.[50] A 'fear-based approach' to deterrence may be counter-productive if it constructs animal protection as an illegitimate goal and a point of community resistance.[51] Further, attempts by national authorities to establish and protect animal habitats have been described as 'green grabbing', or the expropriation of space—particularly from marginalised peoples—for environmental reasons. These strategies may have a neoliberal economic dimension, if private entrepreneurs operate the parks that are supposed to protect animals.[52]

---

[47]See, e.g., Adam, *Elephant Treaties* 2014; Andreas/Nadelmann, *Policing the Globe* 2006, 46–50.

[48]Pires/Moreto, 'The Illegal Wildlife' 2016.

[49]Brown/Cloke, 'The Critical Business of Corruption' 2006, 238; Brown/Cloke, 'Critical Perspectives on Corruption' 2011, 118.

[50]See generally Kelly/Ybarra, 'Introduction to themed issue: "Green security in protected areas"' 2016, 171–175; Humphreys/Smith, 'Militarised Responses to the Illegal Wildlife Trade' 2018, 25–42.

[51]Moreto/Gau, 'Deterrence, Legitimacy, and Wildlife Crime in Protected Area' 2017, 51.

[52]Massé/Lunstrum, 'Accumulation by Securitization' 2016, 227–237.

5   Conclusion

There is thus a need for caution in accepting calls for more and stronger measures to combat wildlife trafficking and corruption. Bribery is seen to enable wildlife trafficking, and wildlife trafficking is seen be one of those illicit industries that motivates active and passive corruption. The two sets of offences can be addressed simultaneously, it is said, and perhaps alongside other efforts to ensure international peace, security, and transnational crime control. However, the effectiveness of economic crime control regimes may be hard to determine and their more draconian mechanisms may generate tensions with individual civil and political rights. Other prospects for fragmentation become clear when special duties to protect collective interests are factored in. Further, the anti-corruption and anti-wildlife trafficking movements may as anthropocentric and/or Eurocentric, especially if they are put together. Finally, I wondered whether a securitised transnational animal law would create new forms of environmental injustice and be self-defeating. Could it fail to protect animals while justifying limits on a range of other public goods?

References

Adam, R. (2014). *Elephant treaties: The Colonial legacy of the biodiversity crisis*. Hanover: University Press of New England.

Andreas, P., & Nadelmann, E. (2006). *Policing the globe: Criminalization and crime control in international relations* (pp. 46–50). Oxford: Oxford University Press.

Ayling, J. (2013). What sustains wildlife crime? Rhino horn trading and the resilience of criminal networks. *Journal of International Wildlife Law & Policy, 16*, 57–80.

Boister, N. (2002). Human rights protections in the suppression conventions. *Human Rights Law Review, 2*, 199–227.

Boister, N. (2003). "Transnational Criminal Law"? *European Journal of International Law, 14*, 953–976.

Bowman, M., Davies, P., & Redgwell, C. (2010). *Lyster's international wildlife law* (2nd ed.). Cambridge: Cambridge University Press.

Brown, E., & Cloke, J. (2006). The critical business of corruption. *Critical Perspectives on International Business, 2*, 275–298.

Brown, E., & Cloke, J. (2011). Critical perspectives on corruption: An overview. *Critical Perspectives on International Business, 7*, 116–124.

Gless, S. (2015). Bird's-eye view and worm's eye view: Towards a defendant-based approach in transnational criminal law. *Transnational Legal Theory, 6*, 117–140.

Gray, L., Hansen, K., Recica-Kirkbride, P., & Mills, L. (2014). (for the International Bank for Reconstruction and Development/World Bank/Organisation for Economic Co-operation and Development), 'Few and Far: The Hard Facts on Stolen Asset Recovery', available at: http://www.oecd.org/dac/accountable-effective-institutions/Hard%20Facts%20Stolen%20Asset%20Recovery.pdf.

Halliday, T., Levi, M., & Reuter, P. (2014). (for the Centre on Law & Globalization), 'Global surveillance of dirty money: Assessing assessments of regimes to control money laundering and combat the financing of terrorism', available at: http://www.americanbarfoundation.org/publications/752.

Heimann, F., Foldes, A., & Coles, S. (2015). (for Transparency International), 'Exporting corruption: Progress report 2015: Assessing enforcement of the OECD convention on combatting foreign bribery', available at: http://www.transparency.org/exporting_corruption.

Humphreys, J., & Smith, M. L. R. (2018). Militarised responses to the illegal wildlife trade. In T. Reitano, S. Jesperson, & L. B. R.-B. de Lugo (Eds.), *Militarised responses to transnational organised crime* (pp. 25–42). Cham: Palgrave Macmillan.

Ivory, R. (2014). *Corruption, asset recovery, and the prevention of property in public international law: The human rights of bad guys*. Cambridge: Cambridge University Press.

Ivory, R. (2017). Asset recovery in four dimensions: Returning wealth to victim countries as a challenge for global governance. In K. Ligeti & M. Simonato (Eds.), *Chasing criminal money: Challenges and perspectives on asset recovery in the EU* (pp. 176–210). Oxford: Hart/ Bloomsbury.

Kelly, A., & Ybarra, M. (2016). Introduction to themed issue: "Green security in protected areas". *Geoforum, 69*, 171–175.

Massé, F., & Lunstrum, E. (2016). Accumulation by securitization: Commercial poaching, neoliberal conservation, and the creation of new wildlife frontiers. *Geoforum, 69*, 227–237.

Moreto, W., & Gau, J. (2017). Deterrence, legitimacy, and wildlife crime in protected areas. In M. Gore (Ed.), *Conservation criminology* (pp. 45–58). Chichester: John Wiley & Sons.

Moreto, W., Brunson, R., & Braga, A. (2015). "Such misconducts don't make a good ranger": Examining law enforcement ranger wrongdoing in Uganda. *British Journal of Criminology, 55*, 359–380.

Peters, A. (2014). *Novel practice of the Security Council: Wildlife poaching and trafficking as a threat to the peace*, EJILTalk!, available at: https://www.ejiltalk.org/novel-practice-of-the-security-council-wildlife-poaching-and-trafficking-as-a-threat-to-the-peace.

Pires, S., & Moreto, W. (2016). The illegal wildlife trade. In *Oxford Handbooks Online*, available at: http://www.oxfordhandbooks.com/view/10.1093/oxfordhb/9780199935383.001.0001/oxfordhb-9780199935383-e-161?print=pdf.

Sharman, J. (2017). *The despot's guide to wealth management: On the international campaign against grand corruption* (pp. 13–14). Ithaca: Cornell University Press.

Smith, R., & Walpole, M. (2005). Should conservationists pay more attention to corruption? *Oryx, 39*, 251–256.

van Uhm, D., & Moreto, W. D. (2017). Corruption within the illegal wildlife trade: A symbiotic and antithetical enterprise. *The British Journal of Criminology, 58*(4), 864–885, available at: https://doi.org/10.1093/bjc/azx032. (advanced publication).

Vettori, B. (2013). Evaluating anti-money laundering policies: Where are we? In B. Unger & D. van der Linde (Eds.), *Research handbook on money laundering* (pp. 474–485). Cheltenham: Edward Elgar.

Wyatt, T. (2013). *Wildlife trafficking: A deconstruction of the crime, the victims and the offenders*. London: Palgrave Macmillan.

Wyatt, T., & Cao, A. N. (2015). Corruption and wildlife trafficking, *U4 Issue, 11*, available at: www.u4.no/publications/corruption-and-wildlife-trafficking/downloadasset/3832.

Radha Ivory (Dr iur) is a Senior Lecturer in Law at the University of Queensland, Australia (UQ). She teaches company law and researches problems at the intersection of criminal law, public international law, and corporate governance. Her monograph, *Corruption, Asset Recovery, and the Protection of Property in Public International Law: The Human Rights of Bad Guys*, appeared with Cambridge University Press in 2014. Her current projects concern the emergence and scope of corporate criminal liability regimes and the transnational legal theory of corruption. Prior to joining UQ, Radha worked in the international sector and in private practice.

Part III
New Legal Concepts

Chapter 9
Biodiversity, Species Protection, and Animal Welfare Under International Law

Guillaume Futhazar

Abstract The chapter explores the influence of the concept of animal welfare on international biodiversity law. A close examination of the recent evolution of this branch of international law shows that animal welfare has an ambivalent place in biodiversity-related agreements. Indeed, while welfare is only a faint consideration in the development of international regimes dealing with biodiversity as a whole, the concept has become an essential element for agreements dealing with the conservation of specific endangered species. Despite its role in these agreements, the place of animal welfare in international biodiversity law highlights that this corpus of rules is currently insufficient to be an effective tool for the protection of wildlife welfare. The last section of this study suggests that the adoption of international rules aiming at ensuring the protection of wild animals' welfare could serve the double purpose of strengthening the conservation purpose of biodiversity regimes while also filling the welfare gap of international biodiversity law.

1 Introduction

In a recent study, researchers from the Weizmann Institute of Science and the California Institute of Technology estimated that humans represent, in terms of mass, only 0.01% of all life.[1] Yet, despite this low 'weight', our collective impact on the biosphere is so significant that we may very well have triggered the sixth mass extinction.[2] This study adds to an already long list of demonstrations as to our impact

---

[1] Bar-On/Phillips/Milo, 'Biomass Distribution' 2018, 1–6.
[2] For an overview of the extent of this phenomenon, see Kolbert, *Sixth Extinction* 2014.

G. Futhazar (✉)
Max Planck Institute for Comparative Public Law and International Law, Heidelberg, Germany
e-mail: futhazar@mpil.de

© The Author(s) 2020
A. Peters (ed.), *Studies in Global Animal Law*, Beiträge zum ausländischen öffentlichen Recht und Völkerrecht 290,
https://doi.org/10.1007/978-3-662-60756-5_9

95

on all living things,[3] and once more calls for a collective reflection on how to mitigate the inexorable human-caused erosion of the earth's biodiversity. Our influence on the environment also raises serious concerns for the living conditions of the remaining surviving life forms that are subject to considerable and sustained pressure.[4] In this context, it becomes essential to examine not only what has been done to cope with this alarming erosion of life but also to deal with the toll we are inflicting on the living organisms that remain.

This chapter aims to explore the influence of the concept of animal welfare in international biodiversity law. To do so, it is necessary, as a preliminary clarification, to define these two terms and illustrate how they relate (Sect. 2). From there, this chapter will highlight the ambivalent place of animal welfare in biodiversity related regimes. Indeed, while the concept of animal welfare is practically invisible in the context of treaties dealing with biodiversity as a whole (Sect. 3), it appears that welfare is gradually becoming a *sine qua non* condition for conservation and sustainable use in the context of international agreements dealing with endangered species (Sect. 4). Based on this analysis, the last section will highlight the relevance of enacting specific international rules to ensure the welfare of wildlife (Sect. 5). Even though these rules might have a distinct purpose from international biodiversity law, they could nevertheless complement it in achieving conservation and sustainable development.

2   The Scope of Animal Welfare and International Biodiversity Law: Wildlife as an Overlapping Theme

The concept of animal welfare generally refers to the living and dying conditions of animals in the context of their different relations with humans. Animal welfare calls for the avoidance of unnecessary suffering and can consequently be understood as the condition in which an animal is free from hunger, malnutrition and thirst; free from fear and distress; free from physical and thermal discomfort; free from pain, injury and disease; and free to express normal patterns of behaviour.[5] Furthermore, animal welfare is related to three overlapping dimensions: the animal's basic health and functioning; its affective state; and its natural way of living.[6] Importantly, it has to be stressed that the concept of welfare makes sense only in the context of

---

[3]For instance, the Global Environmental Outlook by the United Nations Environmental Program (UNEP), or the several Global Biodiversity Outlooks produced in the context of the Convention on Biological Diversity. See *Global Environmental Outlook 5 Environment for the Future We Want* (Nairobi: UNEP 2012); *Global Biodiversity Outlook 4* (Montréal: Secretariat of the Convention on Biological Diversity 2014).

[4]Ibid.

[5]These five 'liberties' were developed by the International Organisation for Animal Health (OIE). For an overview of the OIE's approach to animal welfare, see www.oie.int/en/animal-welfare/animal-welfare-at-a-glance/.

[6]Paquet/ Darimont, 'Wildlife conservation' (2010), 179; Peters, 'Global Animal Law 2016, 11.

human–animal relations.[7] Indeed, an animal in its natural and undisturbed conditions may very well be subjected to events that violate its 'freedoms', such as predation or starvation. The concept of animal welfare does not suggest that such situations should be prevented but rather dictates that humans should not create conditions that negate the aforementioned freedoms.[8]

The influence of the idea of avoiding inflicting unnecessary suffering on animals has grown over the past decades. The increase in academic writing on the subject has been dubbed 'the animal turn'[9] and has fuelled numerous public debates on the subject.[10] Concurrently, states and international organisations have enacted laws and standards aimed at ensuring animal welfare in various settings (domestic animals, agriculture, and scientific experimentation...).[11] Yet, most of these laws and standards concern captive animals, and the welfare of wildlife has been largely ignored.[12] Considering this legal gap at the national and regional levels, the provisions regarding wildlife welfare have to be sought elsewhere and international biodiversity law seems, at first glance, to be suited for this purpose.

International biodiversity law can be understood as the corpus of international rules that aim to ensure the conservation and sustainable use of biodiversity. According to the Convention on Biological Diversity (CBD),[13] biodiversity refers to:

> the variability among living organisms from all sources including, inter alia, terrestrial, marine and other aquatic ecosystems and the ecological complexes of which they are part; this includes diversity within species, between species and of ecosystems.[14]

Biodiversity is an encompassing concept and consequently this branch of international environmental law can take many forms, from agreements dealing with biodiversity as a whole[15] to treaties aiming at preserving a specific ecosystem at the

---

[7]As underlined by numerous commentators, the distinction between humans and animals is misleading since we are also, biologically speaking, animals. It would be more accurate to use the terms 'human animal' and 'non-human animal'. However, for the sake of brevity, this contribution will use the terms 'human' and 'animal', though the exact distinction should be kept in mind.

[8]In sum, welfare is not about eradicating suffering, which can be an integral part of the natural existence of animals. The main logic is to avoid causing additional and unnecessary suffering.

[9]This trend has been the subject of abundant comments. For a brief overview, see Weil, 'Report on the Animal Turn' 2010.

[10]For instance, by revealing footage of slaughter houses, the French association 'L-214' has generated a widely mediatised controversy. See Garric, 'L214, la méthode choc pour dénoncer les abattoirs', Le Monde (29 March 2016).

[11]The 'Global Animal Law' project has compiled a database of all national legislations regarding animal rights and welfare. See www.globalanimallaw.org/database/national/index.html.

[12]For the French legal system, see Nouët, 'L'Animal Sauvage' 2013. In her general study of animal law across countries, Sabine Brels concludes that, currently, free wild animals are the forgotten subjects of animal law. See Brels, Droit du Bien-Être Animal 2017a, 418.

[13]Convention on Biological Diversity (CBD), 5 June 1992, 1760 UNTS 79.

[14]Art. 2. Use of terms.

[15]Mainly the Convention on Biological Diversity (CBD), 1992 (n. 13).

global or regional level[16] or specialised conventions with the purpose of preserving certain endangered species.[17] These latter specialised conventions—which together are sometimes referred to as 'international wildlife law'[18]—stress that species are integral components of the ecosystems in which they live. Their conservation and sustainable use are therefore a prerequisite for the preservation of biodiversity as a whole. Furthermore, several studies have shown that the welfare of the individuals that make up a species is an important aspect for the conservation of the population as a whole.[19] Accordingly, it is possible to infer that animal welfare is taken into account in international biodiversity law since it contributes to the health of species as a whole.

However, as the following sections will show, the place of animal welfare in international biodiversity law is ambiguous. In international agreements dealing with biodiversity in general, animal welfare appears to be a very minor consideration, while in the context of international conventions dealing specifically with the issue of endangered species, animal welfare is gradually becoming an essential element for conservation and sustainable use.

3 Animal Welfare: An Absent Topic in General International Biodiversity Law

The most emblematic biodiversity-related international agreement is arguably the Convention on Biological Diversity (CBD).[20] Despite being weakly normative,[21] this agreement has had a major impact on the general development of international biodiversity law.[22] Several conservation principles that were developed within its

---

[16]For instance, on specific ecosystems, the Convention on Wetlands of International Importance especially as Waterfowl Habitat (Ramsar Convention), 2 February 1971, 996 UNTS 250. Regional instruments are numerous; for marine ecosystems, one could quote the Convention for the Protection of the Mediterranean Sea Against Pollution (Barcelona Convention), 16 February 1976, 1102 UNTS 107; for terrestrial ecosystems, see the Convention on the protection of the Alps (Alps Convention), 6 April 1993, 1917 UNTS 135.

[17]For instance, the Convention on the Conservation of Migratory Species of Wild Animals (CMS), 1 November 1983, 2742 UNTS 197.

[18]For a general overview of this field, see Bowman/Davies/Redgwell, *Lyster's International Wildlife Law* 2010.

[19]The last section of this chapter will further develop this notion.

[20]The CBD was one of the three multilateral agreements adopted following the Rio Summit on the environment in 1992 together with the United Nations Framework Convention on Climate Change and the United Convention to Combat Desertification.

[21]Most of the obligations laid down in the convention are tempered by the following provision: 'as far as possible and as appropriate'.

[22]For an overview, see Morgera/Razzaque (Eds.), *Biodiversity and Nature Protection Law* 2017. Each chapter in this book deals with a specific topic where the influence of the CBD is highlighted (island biodiversity, biosafety, access to genetic resources and so on).

institutions were subsequently adopted by the members of other multilateral environmental agreements (for instance, the ecosystem approach[23] or the Addis Ababa principles on sustainable use of biodiversity[24]), and its current strategic plan and 'Aichi Targets'[25] have become a major reference point for the implementation of the 'biodiversity cluster'.[26] Still, the question of animal welfare is never once addressed in the provisions of the CBD and is, at best, an incidental concern in the corpus of decisions taken by its parties in order to support its implementation.

It can also be said that the heavily anthropocentric and holistic approach that is inherent to the CBD prevents animal welfare considerations to be included in its scope. The development of the ecosystem approach principle is particularly illustrative of this bias.[27] The ecosystem approach calls for appropriate consideration of the entirety of each ecosystem (constituents and processes) in order to ensure that human society benefits from healthy ecosystems.[28] Though this principle does not directly go against the concept of animal welfare, its strong emphasis on human benefits[29] and its holistic approach is illustrative of the opposition that exists between environmental ethics (anthropocentric and holistic) and animal ethics (zoocentric and individualist).[30]

---

[23]CBD, Decision V/6 (May 2000), Ecosystem Approach. This approach had an impact on the conceptualisation of key concepts for the Ramsar Convention. On this topic, see Davison/Coates, 'Ramsar Convention' 2011, 199–205.

[24]CBD, Decision VII/12 (April 2004), Sustainable Use (art. 10). The Addis Ababa Principles comprised in this decision were subsequently adopted by the members of the CITES. CITES, Resolution Conf. 13.2, Rev. Cop.14 (2007), Sustainable use of biodiversity: Addis Ababa Principles and Guidelines.

[25]CBD, Decision X/2 (October 2010), The Strategic Plan for Biodiversity 2011–2020, including the Aichi Biodiversity Targets.

[26]The term 'biodiversity cluster' refers to the main multilateral environmental agreements in the field of biodiversity and ecosystems. These conventions are, in chronological order, the Convention on Wetlands of International Importance Especially as Waterfowl Habitat (Ramsar Convention), 1971 (n. 16); the Convention for the Protection of the World Cultural and Natural Heritage (World Heritage Convention), 16 November 1972, 1037 UNTS 151; Convention of International Trade in Endangered Species of Wild Fauna and Flora (CITES), 3 March 1973, 993 UNTS 243; the Convention on Migratory Species of Wild Animals (CMS), 1983 (n. 17); the Convention on Biological Diversity (CBD), 1992 (n. 13), and the International Treaty on Plant Genetic Resources for Food and Agriculture (ITPGRFA), 3 November 2001, 2400 UNTS 303. For a more detailed account of this 'cluster', see UNEP-WCMC, *Promoting Synergies within the Cluster of Biodiversity-Related Multilateral Environmental Agreements* (Cambridge: UNEP-WCMC 2012). On the diffusion of the Aichi Targets and their legal influence, see Futhazar, 'Diffusion' 2015.

[27]It should be noted that this principle exists in various forms in different international regimes. On the versatility of this notion, see De Lucia, 'Competing Narratives' 2015.

[28]Ibid., 'Ecosystems should be managed for their intrinsic values and for the tangible or intangible benefits for humans, in a fair and equitable way'.

[29]CBD, Decision V/6, 'Ecosystems should be managed for their intrinsic values and *for the tangible or intangible benefits for humans*, in a fair and equitable way' (emphasis added).

[30]To simplify, the main divide between environmental ethics and animal ethics is that environmental ethics consider the environment as a whole and see animals as components of species and not as individuals. On the other hand, animal ethics consider the animal in its individuality and is therefore

Despite this overarching holistic approach to conservation, the decisions of the members of the CBD are not entirely devoid of animal welfare considerations. For instance, art. 8.h of the convention calls for the prevention of the introduction of invasive species and, in cases where this prevention fails, their control or eradication.[31] To implement this article, the parties and institutions of the CBD have developed a series of guidelines that state on several occasions that the eradication of invasive species has to be done in an 'ethically acceptable' way.[32] Of course this mention of ethics is vague but it does offer an entry point for considerations of animal welfare as the topic gains importance at the international level and in national contexts.[33]

Interestingly, the provisions and developments of the CBB concerning the conservation of biodiversity are mirrored in other agreements, be they on specific conservation issues (for instance wetlands[34]) or on specific regions (the Mediterranean[35] or the Alps,[36] for instance). Consequently, as far as agreements concerning biodiversity as a whole are concerned, it is clear that animal welfare is not the main consideration for states or institutions, who are committed instead to a holistic approach for conservation and sustainable use.[37]

Recent developments in other international fora tend to reinforce this assessment. For instance, resolution 72/223 of the United Nations General Assembly on Harmony with Nature clearly highlights the ecosystemic and holistic approach of the UN with regards to the environment.[38] It can also be said that the recent decision of

---

concerned with welfare and rights. On the divide between these approaches, see Sagoff, 'Animal Liberation' 1983; Guichet, 'La question animale' 2013.

[31] Invasive alien species are a serious threat to environmental integrity, even more so than climate change in certain countries. See IPBES, IPBES/2/16/Add.3 (2013), Initial scoping for the thematic assessment of invasive alien species and their control, § 4.

[32] The guidelines can be found in CBD, Decision V/8 (May 2000), Alien species that threaten ecosystems, habitats or species.

[33] The same mention of ethics is also present in the Addis Ababa principles (n. 24).

[34] Davison/Coates, 'The Ramsar Convention' 2011.

[35] Barcelona Convention, Decision IG.17/6 (2008), Implementation of the ecosystem approach to the management of human activities that may affect the Mediterranean marine and coastal environment.

[36] The Protocol on Nature Protection and Landscape Conservation of the Alpine Convention reflects the CBD's orientations. CBD, Memorandum of Cooperation between the Convention on Biological Diversity and the Alpine Convention and the Carpathian Convention (May 2008).

[37] Of course, some counter examples can be mentioned, such as the Antarctic Environmental Protocol which contains explicit references to animal welfare (annex II, art. 3.6). Nevertheless, when it comes to addressing the conservation of biodiversity as a whole, animal welfare is clearly a secondary concern with regards to more holistic approaches.

[38] United Nations, General Assembly, Harmony with Nature, A/RES/72/223, 20 December 2017. This resolution is the latest in a series of resolutions with the same title. The first one was adopted in December 2009 and initiated a process of institutional work in order to promote 'the construction of a new, non-anthropocentric paradigm in which the fundamental basis for right and wrong action concerning the environment is grounded not solely in human concerns'. For more information, see www.harmonywithnatureun.org/.

the ICJ in the *Certain Activities carried out by Nicaragua in the Border Area* case strengthens the legal paradigm of a holistic approach for the conservation of biodiversity.[39] In this historic decision in which the ICJ granted compensation for environmental damage for the first time,[40] the Court decided to evaluate the damage caused to the environment by considering the affected ecosystem in its entirety.[41] When doing so it referred to several concepts: in particular, to the concept of ecosystem services.[42] Ecosystem services refer to the direct and indirect contributions of ecosystems to human well-being. These contributions can be attributed a value in order to determine the economic importance of ecosystems.[43] In this case, by referring to this concept, the ICJ stressed once more the anthropocentric and ecosystemic aspect of international biodiversity law. The recognition of this paradigm for the valuation of biodiversity by the judicial organ of the United Nations is telling of how little the question of animal welfare seems to bear on the evolution of this branch of international law.

However, if animal welfare is but a faint consideration in general international biodiversity law, the diagnosis is quite the opposite when examining agreements concerned with endangered species.

4   Animal Welfare: A Condition for the Sustainable Use and Conservation of Endangered Species

Historically, international environmental law has developed through the adoption of treaties dealing with the conservation and sustainable use of specific species. These treaties had different *raisons d'être*, from the preservation of purely commercial interests[44] to the conservation of species that were deemed useful.[45] As the decades passed, and with the development of scientific knowledge on the global state of the environment and the conservation status of species, several other specific treaties were adopted in order to prevent extinctions. The instruments adopted during the

---

[39]ICJ, *Certain Activities carried out by Nicaragua in the Border Area (Costa Rica v. Nicaragua)*, Judgement of 2 February 2018, ICJ Reports 2018.

[40]Ibid., para. 41.

[41]Ibid., para. 78.

[42]Ibid., para. 52.

[43]This concept is currently at the centre of several international initiatives. It is one of the core concepts of the general framework of the Intergovernmental Platform on Biodiversity and Ecosystem Services (IPBES) and is being promoted by the Economics of Ecosystems and Biodiversity Initiative (TEEB) so as to influence decision-making across the globe. See www.ipbes.net/ and www.teebweb.org/.

[44]Agreement on Measures for Regulating the Catch and Conserving Stocks of Seals in the North-Eastern Part of the Atlantic Ocean, 22 November 1957, 309 UNTS 269.

[45]For instance, the Convention Internationale du 19 mars 1902 pour les Oiseaux Utiles à l'Agriculture, available at: www.admin.ch/opc/fr/classified-compilation/19020002/index.html.

second half of the twentieth century no longer only dealt with specific species but also with practices such as international trade or with broader categories, such as migratory species. Yet, with the exception of the well-known example of the CITES,[46] these different instruments make no mention of the concept of welfare, be it directly or indirectly, in their provisions. For instance, the GORILLA Agreement,[47] adopted within the context of the Convention on Migratory Species, calls for the maintenance of gorillas 'in a favourable conservation status'.[48] The same goes for the Agreement on the Conservation of Polar Bears,[49] which states that its parties 'shall manage the bear population in accordance with sound conservation practice'.[50] Other treaties generally prohibit the killing of individual members of certain species[51] but do not provide any other indication on how to ensure that the same individuals are not subjected to unnecessary suffering due to restrictions to their animal freedoms by human activities.

At first glance, it appears that the welfare of individual members of those specific species concerned by these numerous instruments is not considered essential for their conservation. However, most international environmental treaties have an evolutive purpose by design[52] and their main provisions are often followed by decisions and guidelines[53] that allow for an updated implementation in order to achieve greater effectiveness. This process of legal densification[54] of the primary obligations of states with regard to the conservation of certain species has paved the way for considerations of welfare. For instance, with regards to the conservation of cetaceans, the influence of welfare on the obligations of states is increasingly clear. This is most notable in the context of the International Whaling Commission, where its members have adopted several decisions aiming at ensuring the welfare of

---

[46]Convention of International Trade in Endangered Species of Wild Fauna and Flora (CITES), 1973 (n. 26). See, inter alia, art. III.2.c, 'An export permit shall only be granted when the following conditions have been met (. . .): a Management Authority of the State of export is satisfied that any living specimen will be so prepared and shipped as *to minimize the risk of injury, damage to health or cruel treatment*' (emphasis added).

[47]Agreement on the Conservation of Gorillas and their Habitats, 26 October 2007, 2545 UNTS 55.

[48]Art. II.

[49]Agreement on the Conservation of Polar Bears, 15 November 1973, 2898 UNTS I-50540.

[50]Art. II.

[51]Such as the agreements and memoranda of the 'CMS family', i.e. all the international agreements, formal or informal, adopted in order to implement the CMS with regards to specific migratory species. For an overview of the formal agreements, see www.cms.int/en/cms-instruments/agreements.

[52]As underlined by the ICJ. ICJ, *Case Concerning the Gabčikovo-Nagymaros Project (Hungary v. Slovakia)*, Judgement of 25 September 1997, ICJ Reports 1997, 7, para. 104.

[53]Mostly decisions by the meeting of the parties held on a regular basis (once every two to three years).

[54]On this topic, see Churchill/Ulfstein, 'Autonomous Institutional Arrangements' 2000; Brunnée, 'COPing with Consent' 2002; Wiersema, 'New International Law-Makers?' 2009.

cetaceans falling under the scope of the commission.[55] Moreover, several decisions taken in other international fora encourage states to take into consideration the cultures of whale populations in their conservation effort.[56] Similar developments focusing on welfare can be seen in the context of other species-related treaties. The parties to the EUROBATS agreements[57] have adopted decisions specifying how injured individual bats should be taken care of before being released into the wild.[58] In the context of the Agreement on the Conservation of Albatrosses and Petrels (ACAP),[59] several technical guidelines, which can be used as interpretive tools for the general provisions of the agreement,[60] were produced to ensure the welfare of individuals (e.g. the guidelines on translocation).[61]

Witnessing the evolution of international treaties dealing with the conservation and sustainable use of specific species, one gains the impression that the welfare of individuals is gradually becoming a prerequisite for the conservation of species as a whole. As the knowledge of species and the pressures they endure grows, it is apparent that ensuring their conservation cannot simply be confined to the strict application of the main provisions of international agreements. Not killing, or killing within commonly agreed limits (for instance, on the basis of quotas), is insufficient on its own to preserve the specie, and so is 'simply' preserving habitats. Active steps must be taken by the parties to these international instruments in order to guarantee collective welfare and thus effectively achieve conservation and sustainable use.

However, there are two limits to this diagnosis that seem to equate animal welfare and international biodiversity law. Firstly, the provisions concerning welfare are mostly contained in secondary rules or technical guidelines. As such, they are not binding. This leaves considerable leeway to the states in enforcing them. Secondly— and this is the greatest limitation—even though the influence of animal welfare in the evolution of these specific instruments is clear, they only concern a very small proportion of wildlife, namely species that have been recognised as endangered and that are the object of international rules. Solely based on these agreements, it would be a gross overstatement to claim that the welfare of wildlife is guaranteed by international biodiversity law.

---

[55]This aspect has been abundantly commented upon. See, for instance, Harrop, 'From Cartel to Conservation', 2003.

[56]CMS, Résolution 11.23 (2014), Conservation Implications of Cetacean Culture, para. 2. The consideration for the culture of Cetaceans has subsequently been added to decisions of other cetacean-related instruments, such as the ACCOBAMS in the Mediterranean region: ACCOBAMS, Resolution 6.14 (2016), Population Structure Studies (the resolution cites the CMS resolution in its opening considerations).

[57]Agreement on the Conservation of Bats in Europe (EUROBATS), 4 December 2012, 1863 UNTS 101.

[58]EUROBATS, Résolution 7.10 (2014), Bat Rescue and Rehabilitation.

[59]Agreement on the Conservation of Albatrosses and Petrels, 19 June 2001, 2258 UNTS 257.

[60]As provisioned by art. 32 (complementary means of interpretation) of the Vienna Convention on the Law of Treaties, 23 May 1969, 1155 UNTS 331.

[61]Jacobs/Deguchi/Perriman et al., *Guidelines* 2015, available at: https://www.acap.aq/en/resources/acap-conservation-guidelines/2640-translocation-guidelines/file.

Hence, it can be said that international biodiversity law, in its general and specific aspects, does not constitute an appropriate tool to ensure the welfare of wildlife in its entirety. International biodiversity law concerns itself with animal welfare only when it deals with the conservation and sustainable use of specific species or the regulation of certain practices, as welfare becomes a condition for the achievement of its purpose.[62] Consequently, this welfare gap in international biodiversity law means that specific rules would have to be enacted in order for the welfare of wildlife to be protected at the international level. The following section will show that such rules, though they may have a purpose that is distinct from conservation or sustainable use, could complement international biodiversity law.

5   The Relevance of Complementary International Rules to Ensure Both Wildlife Welfare and Conservation

It is important to stress that while welfare and conservation do not have the same ethical basis (individualist versus collective) they might lead to similar practices and end results.[63] This section will examine whether a specific set of rules concerning animal welfare at the global level could fill the gaps in international biodiversity law and complement its existing dispositions.

Several authors have called for the adoption of international instruments and/or rules specifically designed to ensure the protection of animals as individuals.[64] In doing so, they argue that the lack of rules that could guarantee the welfare of wildlife is morally unsatisfactory with regard to the unnecessary suffering that human activities cause to individual animals, both captive and wild.[65] For instance, in an article published in 2012, David Favre argued in favour of the adoption of a framework convention that would establish common principles regarding the welfare of animals at the international level.[66] Such a convention would contain articles

---

[62]Stuart Harrop has suggested that as biodiversity declines, welfare considerations will become increasingly predominant in international environmental law. The rationale is that the few remaining wild animals will have to be preserved by fully taking into account all the elements necessary for their welfare: Harrop, 'Climate Change' 2011.

[63]For instance, the question of traps and their indiscriminate and cruel effect on wildlife is simultaneously a concern for conservationists and welfarists alike. On the interplay between welfare and conservation with regards to cetaceans, see Harrop, 'From Cartel to Conservation' 2003.

[64]For instance, Favre, 'An International Treaty' 2012; Brels, 'A Global Approach', 2017b.

[65]Adam/Schaffner, 'International Law and Wildlife Well-Being' 2017; Scholtz, 'Injecting Compassion' 2017.

[66]Favre, 'An International Treaty' 2012.

calling for a reduction of the killing and unnecessary suffering of wildlife, as well as the preservation of habitats.

In addition to the ethical arguments, it is also relevant to underline the fact that welfare considerations can be useful for the conservation of populations, thus setting the plea for the welfare of wildlife in the context of environmental ethics.[67] For instance, several studies have demonstrated how high stress levels in individuals can affect the overall population. This factor can influence the success rates of species reintroduction practices[68] or aggravate the spread of diseases within a population.[69] At the international level, the ongoing effort to mitigate the adverse impact of anthropogenic noise on cetaceans is a clear illustration that welfare and conservation can be intertwined.[70]

This goes to show that enacting international rules for the welfare of animals, and *ipso facto* wildlife, could serve the double purpose of filling the ethical gap with regard to the impact of human activities on animals and strengthening existing conservation regimes. Rules concerned with animals as individuals can complement rules concerned with animals as species.[71] The last question then is to determine what form such rules would take.[72]

As mentioned earlier, several authors have called for an international treaty that would exclusively deal with the question of welfare and animal rights.[73] An alternative to a stand-alone treaty would be to insert elements of welfare into existing or future environmental agreements. For instance, the UNGA has recently adopted a resolution launching a process that could eventually lead to the adoption of a Global Environment Pact.[74] Such a pact could possibly call for the due consideration of

---

[67]Paquet/Darimont, 'Wildlife conservation and animal welfare' 2007, 179: 'The integrity of habitats and the populations they contain are inextricably linked to the welfare of the individual animals that constitute this population and occupy those habitats'.

[68]Hing et al., 'Relationship between physiological stress and wildlife disease' 2016.

[69]Teixeira et al., 'Revisiting translocation' 2007.

[70]Anthropogenic noise and the stress it generates for individual cetaceans has been linked to higher mortality rates and stranding. See, inter alia, Thorne/Johnston, 'Response of Cetaceans' 2007. This topic raises complex legal question on how to reduce and mitigate their adverse effects. See Scott, 'International Regulation of Undersea Noise' 2004; Firestone/Jarvis, 'Response and Responsibility 2007.

[71]Though it must be stressed that it is not systematically the case. For instance, welfare rules could prohibit the use of traps or practices that are particularly cruel but do not have a significant impact on the conservation of the population as a whole.

[72]Discussing their exact content falls outside of the scope of this chapter. For references, see n. 64 above.

[73]See n. 64 above.

[74]United Nations, General Assembly, Towards a Global Pact for the Environment, A/72/L.51, 7 May 2018. A first draft of the pact has been elaborated by a French think tank ('le club des juristes') and is available at: http://pactenvironment.org/fr/.

animal welfare in environmental conservation, especially considering the fact that a growing number of states have enacted laws with regard to the protection of animal wellbeing.[75] For instance, an additional line calling for the prevention of unnecessary suffering could be added to the current art. 2 (duty to take care of the environment) which states that:

> Every State or international institution, every person, natural or legal, public or private, has the duty to take care of the environment. To this end, everyone contributes at their own levels to the conservation, protection and restoration of the integrity of the Earth's ecosystem.[76]

Consideration for animal welfare could also be added to the normative corpus of existing regimes. For instance, the CBD's Aichi Targets will be renewed in 2020.[77] Putting consideration of animal welfare into the renewed targets could ensure that all biodiversity related conventions are implemented not only to ensure conservation but also welfare. Indeed, these Targets have been widely adopted by other multilateral environmental agreements and now constitute commonly shared goals.[78] One can expect the next set of targets to also be included in the normative framework of other biodiversity-related regimes.

These are, of course, optimistic suggestions. Regrettably, international environmental law is largely ineffective, and adding considerations of animal welfare to it would not necessarily lead to an improvement. However, it would constitute a first step that could complement existing national and regional initiatives and encourage states to consider wildlife in their existing animal welfare laws. The most difficult questions lie in the necessary actions that need to be taken in order to move from ideas to concrete and effective realisations.

References

Adam, R., & Schaffner, J. (2017). International law and wildlife well-being: Moving from theory to action. *Journal of International Wildlife Law and Policy, 20*, 1–17.

Bar-On, Y. M., Phillips, R., & Milo, R. (2018, June 19). The biomass distribution on earth. *Proceedings of the National Academy of Science, 115*(25), 6506–6511.

Bowman, M., Davies, P., & Redgwell, C. (2010). *Lyster's international wildlife law* (2nd ed.). Cambridge: CUP.

Brels, S. (2017a). A global approach to animal protection. *Journal of International Wildlife Law and Policy, 20*, 105–123.

[75]The ubiquity of animal welfare provisions in national legislations has led some authors to ponder whether or not animal welfare could be considered a principle of international law. See Sykes, 'Nations Like Unto Yourselves' 2011.

[76]See n. 74 above.

[77]In 2020, the current strategic plan of the CBD will be terminated and replaced with another that will take into consideration the successes and failures of its predecessor. The current plan was adopted using this process when it was recognised that the plan adopted in 2002 had not produced the expected results.

[78]Futhazar, 'The Diffusion of the Strategic Plan for Biodiversity' 2015.

Brels, S. (2017b). *Le Droit du Bien-Être Animal dans le Monde: Evolution et Universalisation*. Paris: L'Harmattan.

Brunnée, J. (2002). Coping with consent: Law making under multilateral environmental agreements. *Leiden Journal of International Law, 15*, 1–52.

Churchill, R., & Ulfstein, G. (2000). Autonomous institutional arrangements in multilateral environmental agreements: A little-noticed phenomenon in international law. *American Journal of International Law, 94*, 623–659.

Davison, N., & Coates, D. (2011). The Ramsar Convention and synergies for operationalizing the convention on biological diversity's ecosystem approach for wetland conservation and wise use. *Journal of International Wildlife Law and Policy, 14*, 199–205.

De Lucia, V. (2015). Competing narratives and complex genealogies: The ecosystem approach in international environmental law. *Journal of Environmental Law, 27*, 91–117.

Favre, D. (2012). An international treaty for animal welfare. *Animal Law, 18*, 237–280.

Firestone, J., & Jarvis, C. (2007). Response and responsibility: Regulating noise pollution in the marine environment. *Journal of International Wildlife Law and Policy, 10*, 109–152.

Futhazar, G. (2015). The diffusion of the strategic plan for biodiversity and its Aichi biodiversity targets within the biodiversity cluster: An illustration of current trends in the global governance of biodiversity and ecosystems. *Yearbook of International Environmental Law, 25*, 133–166.

Garric, A. (2016, March 29). L214, la méthode choc pour dénoncer les abattoirs. *Le Monde*.

Guichet, J.-L. (2013). La question animale dans l'éthique environnementaliste. *Journal International de Bioéthique, 24*, 29–38.

Harrop, S. (2003). From Cartel to Conservation and on to compassion: Animal welfare and the international whaling commission. *Journal of International Wildlife Law and Policy, 6*, 79–104.

Harrop, S. (2011). Climate change, conservation and the place for wild animal welfare in international law. *Journal of Environmental Law, 23*, 441–462.

Hing, S., Narayan, E. J., Thompson, R. C. A., et al. (2016). The relationship between physiological stress and wildlife disease: Consequences for health and conservation. *Wildlife Research, 43*, 51–60.

Jacobs, J., Deguchi, T., Perriman, L., Flint, E., Hummer, H., & Uhart, M. (2015) Guidelines for translocations of albatrosses and petrels (ACAP Secretariat: Macquarie 2015), available at: https://www.acap.aq/en/resources/acap-conservation-guidelines/2640-translocation-guidelines/file.

Kolbert, E. (2014). *The sixth extinction: An unnatural history*. London: Henry Holt and Co.

Morgera, E., & Razzaque, J. (2017). *Biodiversity and nature protection law*. Cheltenham: Edward Elgar.

Nouët, J.-C. (2013). L'Animal Sauvage au regard du droit et de l'éthique en France. *Journal International de Bioéthique, 24*, 65–76.

Paquet, P. C., & Darimont, C. T. (2010). Wildlife conservation and animal welfare: Two sides of the same coin. *Animal Welfare, 19*, 177–190.

Peters, A. (2016). Global animal law: What it is and why we need it. *Transnational Environmental Law, 5*, 9–23.

Sagoff, M. (1983). Animal Liberation and environmental ethics: Bad marriage, quick divorce. *Osgoode Hall Law Journal, 22*, 297–307.

Scholtz, W. (2017). Injecting compassion into international wildlife law: From conservation to protection? *Transnational Environmental Law, 6*, 463–483.

Scott, K. (2004). International regulation of Undersea noise. *International and Comparative Law Quarterly, 24*, 287–392.

Sykes, K. (2011). Nations like unto yourselves: An inquiry into the status of a general principle of international law on animal welfare. *Canadian Yearbook of International Law, 49*, 3–49.

Teixeira, C., Schetini de Azevedo, C., Mendl, M., Cpreste, C. F., et al. (2007). Revisiting translocation and reintroduction programmes: The importance of considering stress. *Animal Behaviour, 73*, 1–13.

Thorne, L. H., & Johnston, D. W. (2007). Response of cetaceans to anthropogenic noise. *Mammal Review, 37*, 81–115.
Weil, K. (2010). A report on the animal turn. *Differences, 21*, 1–23.
Wiersema, A. (2009). The new international law-makers? Conferences of the parties to multilateral environmental agreements. *Michigan Journal of International Law, 31*, 231–302.

Guillaume Futhazar is a senior research fellow at the Max Planck Institute for Comparative Public Law and International Law. He holds a Ph.D. in Public Law from Aix-Marseille University. His current research deals with, *inter alia*, international environmental law, the interrelations between law and science, and regime interactions.

Chapter 10
Toward International Animal Rights

Anne Peters

Abstract The chapter starts from the observation that while animal welfare is increasingly protected in domestic jurisdictions, animal rights are still hardly recognised, although they would serve animals better. It argues that animal rights would need to be universalised in order to deploy effects in a globalised setting. The international legal order is flexible and receptive to non-human personhood which goes with rights. Also, the historical experience with international human rights encourages the animal rights project, because it shows how similar conceptual and normative difficulties have been overcome. Animal rights would complement human rights not the least because the entrenchment of the species hierarchy as manifest in the denial of animal rights in the extreme case condones disrespect for the rights of humans themselves.

1   Introduction: The Spectre of Dehumanisation

In May 2018, US President Donald Trump spoke about illegal border crossings: 'We have people coming into the country, or trying to come in—and we're stopping a lot of them—but we're taking people out of the country. You wouldn't believe how bad these people are. *These aren't people. These are animals.*'[1]

Such *dehumanisation* (in this case: of foreigners at the Californian-Mexican border) has—throughout history—been a standard discursive strategy to prepare, instigate, facilitate, and exculpate violence committed by humans against other humans.[2] It is exactly in reaction to excesses of such dehumanising mass violence

---

[1]The White House, Remarks by President Trump at a California Sanctuary State Roundtable, 16 May 2018, available at: https://www.whitehouse.gov/briefings-statements/remarks-president-trump-california-sanctuary-state-roundtable/ (emphasis added).

[2]Bain/Vaes/Leyens, *Humanness and Dehumanization* 2014.

A. Peters (✉)
Max Planck Institute for Comparative Public Law and International Law, Heidelberg, Germany
e-mail: apeters-office@mpil.de

© The Author(s) 2020
A. Peters (ed.), *Studies in Global Animal Law*, Beiträge zum ausländischen öffentlichen Recht und Völkerrecht 290,
https://doi.org/10.1007/978-3-662-60756-5_10

committed in the Third *Reich* and during World War II that the Universal Declaration of Human Rights (UDHR) was adopted in 1948. This chapter argues that the objectives of the UDHR itself would be furthered if the United Nations (or another international body such as the FAO, the WHO, or the Animal Health Organisation (OIE)) seriously engaged in work on a universal animal rights' declaration. Importantly, the declaration should—firstly—endorse rights (as opposed to welfare). Secondly, it should proclaim universal rights as opposed to rights on the state level. Thirdly, in order to eventually become hard law, it must be backed by governments, not 'only' by civil society organisations, although these need to be involved in its preparation.

While animal welfare is increasingly protected in domestic jurisdictions, animal rights are still hardly recognised, although they would serve animals better (Sect. 1). Animal rights would need to be universalised in order to deploy effects in a globalised setting (Sect. 2). The international legal order is flexible and receptive to non-human personhood (Sect. 3). The historical experience with international human rights encourages the animal rights project, because it shows how similar conceptual and normative difficulties have been overcome (Sects. 4–6). Animal rights would complement human rights not the least because the entrenchment of the species-hierarchy as manifest in the denial of animal rights in the extreme case condones disrespect for the rights of humans themselves (Sect. 7).

2 The Trend Towards Animal Welfare and Rights in Domestic Laws

Since 1948, the rise and entrenchment of human rights in state constitutions and in the international system has not been paralleled by widespread and firm recognition of animal rights. Rather, animals have been protected by objective standards rather than through rights in a growing number of states around the world.[3]

Only very recently, some few domestic jurisdictions have begun to acknowledge animal rights. Courts in Argentina and Colombia have granted habeas corpus to apes[4] and a bear.[5] The Indian Supreme Court recognised fundamental animal rights

---

[3]See two databases: Sabine Brels/Antoine F. Goetschel, 'Animal Legislations in the World at National Level' (status of 1 March 2017), available at: https://www.globalanimallaw.org/data base/national/index.html; World Animal Protection, 'Animal Protection Index', 2014, available at: https://api.worldanimalprotection.org/.

[4]Argentina: Tercer Juzgado de Garantías Mendoza, case no. P-72.254/15, 3 November 2016 - *Chimpanzee Cecilia.*

[5]Colombian Supreme Court of Justice, AHC4806-2017, Radicación n.o 17001-22-13-000-2017-00468-02, 26 July 2017 - *bear Chucho*. This decision was overturned by the Constitutional Court with a public hearing on 8 August 2019.

under the Indian constitution.[6] In a criminal trial against animal activists for trespassing, a German lower court has accepted self-defence in favour of farmed animals which could be creatively read as implying that these are 'persons' within the meaning of the law.[7] The pattern has not been one of unambiguous progress towards recognition of animal rights, however: in US American lower courts, judges have been hesitant to endorse animal rights. Rather, they have denied habeas corpus to chimpanzees[8] and the standing of a Macaque in a copyright suit.[9]

The novel judicial practice in some few states towards animal rights is good for animals themselves, because rights confer a stronger and more sustainable legal protection for the interests of animal individuals than the safeguards offered by 'objective' laws.[10] The benefits of granting (or acknowledging) legal rights to an actor are procedural, legal, social, and symbolic. Rights facilitate standing in court, rights trigger an obligation to justify their curtailment, rights are dynamic with regard to their exact content and addressees, and rights allow adapting the law to evolving moral attitudes. Not all of these blessings of rights are equally relevant for animals. The main asset for animals is that rights confer a legal position which is elevated above the ordinary balancing of conflicting goods. When animals only benefit from protective rules, their welfare is but one interest among others. Balancing the animals' interests against human interests typically ends up prioritising the human interests, even trivial ones. Arguably, this type of balancing is structurally biased against the animals. In contrast, animal rights would allow a fair balancing in which the proper value of fundamental animal interests (such as the interest to live) could be integrated.[11] Animal rights would not categorically rule out animals being slaughtered for food, kept as pets, and used in scientific experiments but they would place a higher burden on the justification of such uses. Animal rights would thus preclude the current routine sacrifice of fundamental animal interests in favour of trite human interests.

---

[6]Supreme Court of India, *Animal Welfare Board of India v. Nagaraja and others*, Civil appeal no. 5387, 7 May 2014. See most recently also High Court of Punjab and Haryana at Chandighar, CRR 533-2013, judgment of 31 May 2019, esp. para. 29: 'The entire animal kingdom including avian and aquatic are declared legal entities having a distinct personality with corresponding rights, duties and liabilities of a living person.'

[7]LG Magdeburg, Az. 28 Ns 182 Js 32201/14 (74/17), 28 Ns 74/17, judgment of 11 October 2017; OLG Naumburg (Saale), judgment of 9 May 2018 rejected self-defence but not because the pigs were not 'others', but rather because the act of trespass did not avert an imminent danger for those same pigs.

[8]See on the various decisions in the state of New York Söhner, 'Habeas Corpus-Beschwerden' 2016.

[9]US Court of Appeals (9th Circ), *Naruto v. Slater*, No. 16-15469 D.C. No.3:15-cv-04324-WHO, judgment of 23 April 2018.

[10]Seminally Stucki, *Grundrechte für Tiere* 2016, notably 296-301.

[11]See in detail Peters, 'Liberté, Égalité, Animalité' 2016, 25-53, sect. 5.

3   The Need for International Animal Rights

But purely domestic animal rights would not be enough. Nation-state based regulation would not suffice because the problem is a global one. The industrialised mode of meat, dairy, and pet production is now spreading to countries of the global south and to developing countries in which the demand and purchase power for animal products is steeply rising. Those industries have become globalised, with transnational supply chains. The manufacturing and trade conditions are leading to a dramatic increase of 'normal' violence against animals in sheer numbers (confinement, mutilation, killing). Also, the transnational dimension of these more or less violent activities has been intensified, it has become cross-border violence.

I submit that the principled arguments which have led to the codification of human rights in international catalogues are relevant for potential animal rights as well. Firstly, from the perspective of fairness and justice, such rights (once accepted as a matter of principle) are incumbent on animals independent of their place of birth and abode, and they are therefore universal. Secondly, international rights would serve as a benchmark for domestic law. International instruments would potentially allow for some monitoring or at least facilitate the formulation of criticism against domestic practices which do not satisfy the international standard. Thirdly, while the main mechanism for enforcing rights in domestic law is a court process where standing for animals creates additional problems, international rights are mainly monitored in non-adversarial reporting procedures in which the rights-holders do not act as parties. The factual difference between human victims and animal victims which cannot speak for themselves does not bear on these proceedings.[12]

Fourthly and most importantly, the endorsement of animal rights only on the national level in some states would probably lead to the outsourcing of the relevant industries.[13] This risk is already present when one state has higher protective standards than others,[14] and it could be exacerbated when one but not all states embrace a rights-based approach to animal protection. In order to prevent a competitive disadvantage of industries subject to higher domestic standards, and in order to forestall a race to the bottom, harmonised universal standards and a level playing field must be sought.[15] Such harmonisation is also desirable to accommodate consumers' concerns about the importation of animal products from low-standard countries, and would obviate import prohibitions based on such public morality concerns.[16]

---

[12]I thank Tom Sparks for making this point.

[13]See also Anne Peters, Introduction, in this volume.

[14]See, e.g., on the outsourcing of animal experiments: Sueur, 'La fuite de la recherche biomédicale' 2016, 19.

[15]Baldwin/Cave/Lodge, *Understanding Regulation* 2012, 362 et seq.

[16]For example, the production of foie gras and frog legs is prohibited as animal cruelty in Switzerland but the import of such products is allowed. Swiss animal organisations have so far in vain sought to introduce an import ban on 'cruelty products'.

This consideration is not alien in the field of human rights, and manifests the parallel urgency. The economic motive to create an obstacle to any attempt at 'social dumping' has at times provided the stimulus for new transnational or supranational regulation spanning various (competing) national economies. The classic example is the provision on equal pay for male and female workers which was inserted in the Treaty on the European Economic Community (now art. 157 TFEU) chiefly to prevent that 'competition is not distorted' by low wages.[17]

For all these reasons, an inter-governmental universal declaration on animal rights is warranted. A different strategy, alternative to writing a separate instrument for animals, would be a novel expansive interpretation of the relevant human rights instruments to extend their application to non-human animals. Along that line, it has been suggested that the terms 'everyone' (such as in Art. 2 of the UDHR), or 'individual' (such as in Art. 2(1) ICCPR) could be re-read so as to encompass animals too.[18] But this 'ecological interpretation' runs against the history and wording of the UDHR which at other places specifically refers to the 'human person' (Preamble of the UDHR) and to 'human beings' (Art. 1 UDHR). More importantly, many of the rights of the UDHR are not relevant for animals or would have to be adapted in order to fit. While animals do not need free speech, freedom of religion, or equal access to public office, sentient animals do need a right to life, a right to be free from torture, and physical liberty[19]—which could be acknowledged in a separate international instrument.

4   International Animal Personhood

In law, personhood (personality, syn. subjecthood) is a precondition or correlate for holding rights. Personhood is best understood as a cluster concept that does not depend on a set of definite properties but has blurry boundaries.[20] The *legal* ascription of personhood is internal to a given legal order.[21] This means that an actor or an entity can be a person for some purposes (or in some subfields of the law) and a nonperson for others.

Importantly, international law has dynamically recognised the personhood of a host of actors, and international law is particularly *open* to the personhood of non-humans—with states being the main persons in this legal order. Humans were

---

[17]ECJ, *Defrenne v. Belgium*, case 80/70, Opinion of Advocate General Dutheillet de Lamothe, ECR 1971, 445 (455).

[18]Fischer-Lescano, 'Natur als Rechtsperson' 2018, 215. In favour of applying the extant fundamental rights provisions to animals, see also Stucki, *Grundrechte* 2016, 352.

[19]In addition, the right to legal personality, the 'right to have rights' would be the explicit or implied pre-condition for all other rights.

[20]Kurki, 'Why Things Can Hold Rights' 2017, 69-89.

[21]Radbruch, 'Rechtsphilosophie', 3rd ed. 1932, 1993, 363-365.

in the late nineteenth century and early twentieth century explicitly and adamantly qualified as 'objects' of international law.[22] Accordingly, early international treaties to suppress the trade in women and girls (often referred to as the 'white slave trade') were intended to preserve morality; rights of women and children were unknown.[23] With regard to animals, that line of reasoning persists. Until the beginning of the twentieth century, all normative restrictions on animal abuse served to protect public morality, 'decency', or 'chastity'. Animal cruelty was a 'public misdemeanour' and prohibited only if it took place in public.

The parallels between the past status of humans in international law and the present status of animals is striking, as a textbook recognises: 'In modern systems of municipal law all individuals have legal personality, but in former times slaves had no legal personality; they were simply items of property. Companies also have legal personality, but animals do not (. . .). In the nineteenth century (. . .) international law regarded individuals in much the same way as municipal law regards animals.'[24]

In international regulation against human trafficking, the purely other-regarding 'public morals'-rationale has been overcome. Similarly, modern domestic animal laws protect animals for their own sake, as sentient beings. However, the difference remains that animals are mostly protected without granting them rights. Only for humans has an actual rights revolution taken place in national and international law.[25]

Against the background that corporations can be persons for purposes of domestic commercial law, and that the legal status of humans has changed from objects to subjects of international law, there is no intrinsic conceptual barrier to assigning international legal personality to animals—basically because personhood is a purely technical juridic device, a legal fiction. Hence, a leap from the protection of animals by rebound to protection through international animal legal rights is legally possible. But in social and cultural terms, this will be a long shot.

5 Animal Rights and Human Rights: Foundations

Scepticism against international animal rights is tempered by recalling that fundamental objections against the internationalisation of rights have likewise bedevilled the international *human* rights regime. In the context of human rights, these problems were, if not resolved then somehow circumvented or brought to productive use.

---

[22]Triepel, *Völkerrecht und Landesrecht* 1899, 20-21. See also Manner, 'Object Theory' 1952.

[23]International Agreement for the Suppression of the White Slave Traffic, 18 May 1904, 1 LNTS 83, Preamble and art. 1; International Convention for the Suppression of the White Slave Traffic, 4 May 1910, 211 Consol. TS 45, art. 2; International Convention for the Suppression of the Traffic in Women of Full Age, 11 October 1933, 53 UNTS 49, art. 1.

[24]Malanczuk, *Akehurst's Modern Introduction* 1997, 91.

[25]See for doubts about the advent of a time-lagged animal rights revolution: Pinker, *Better Angels* 2011, 473-474.

The first problem is that—notwithstanding the entrenchment of international human rights in international hard law texts—the moral, political, and juridico-theoretical value of human rights remains precarious and in endless dispute. The contemporary debate moves away from naturalistic justifications of human rights based on controversial assumptions about human nature.[26] It rather pivots around instrumental justifications ranging from enabling the realisation of capabilities[27] over the protection against vulnerability[28] to the structuration of relationships of power, responsibility, trust, and obligation.[29]

Protagonists of the so-called political approach ('human rights without foundation'[30]) have even given up the search for a normative justification and contend themselves in a purely positivist manner to observe the international human rights practice so as to glean from this to what extent human rights are in fact accepted as an argument that disables the sovereignty-based defence against any outside critique.

Despite these debates and doubts, the oppressed and marginalised of the world seem to regard rights as a useful legal instrument, and continue to reclaim them. This shows that deep theoretical controversy over rights as a legal institution,[31] and specifically over fundamental human rights (both on the national and the international level)[32] has so far not led to the abandonment of rights as a practical institution of positive law—quite to the contrary. And this in turn implies that the academic controversy about animal rights need not be an obstacle for trying them out as a tool for protecting animal interests.

6 Animal Rights and Human Rights: Universality

Both international human rights and potential international animal rights protection face the critique of cultural imperialism. Mirroring the critique against the 'Western' human rights movement, it has been said that the animal rights movement—like the human rights movement before it—is 'yet another crusade by the West against the practices of the rest of the world',[33] and that the propagators of such crusades claim universal validity in order to impose their own purely local preferences on other cultures, so as to consolidate cultural and political dominance over the non-Western

---

[26]Cruft/Liao/Renzo, 'Philosophical Foundations' 2015, 1-41.

[27]Nussbaum, *Frontiers of Justice* 2006.

[28]Albertson Fineman, 'The Vulnerable Subject' 2008.

[29]Nedelsky, *Law's Relations* 2011.

[30]Raz, 'Human Rights without Foundations' 2011, 321-337, especially at 332, building on John Rawls.

[31]Seminally Tushnet, 'An Essay on Rights' 1984.

[32]Dembour, 'Critiques' 2017, 41-59.

[33]This is how the fictional character Thomas O'Hearn, 'professor of philosophy of Appleton', puts it (Coetzee, *Lives of Animals* 1999, 60).

world, especially the Global South. This charge is not trivial. There is a real risk that the protection of animals targets minority practices (such as Muslim ritual slaughter[34] or indigenous seal and whale hunting), although these practices are in numerical terms insignificant in comparison to the majority's 'normal' massive use and killing of animals. This targeting manifests and fuels majority prejudices against the singled out groups, and can pave the way for intervention and domination. In fact, 'dominant groups have long justified their exercise of power over minorities or indigenous peoples by appealing to the "backward" or "barbaric" way they treat (...) animals.'[35]

However, references to cultural traditions suffer from three flaws. First, historical experience shows their frequently pretextual character. Typically, ruling elites abusively invoke 'culture' in order to secure illegitimate privileges. Second, we should not exaggerate cultural difference. The massive use of animals for human needs and the paucity of reflection on and justification of these practices in ethical terms is a shared feature of all cultures. Thirdly, cultures do not unfold inevitably, as if according to a genetically defined pattern. Eating shark soup made from fins cut off live sharks, fox hunting with hounds, staging bullfights, and stuffing geese for foie gras may be a tradition just like relegating women to the house and prohibiting them from exercising certain professions or driving a car. But simply because these are traditions they are not immutable, and are not worth protecting as such. Instead, morals, traditions, and legal provisions (in short: culture) are made, practiced, and applied by human beings capable of learning, and can change.

7 Conclusions

The legal correspondence (and arguably mutual enrichment) of rights for human and non-human animals was intuitive when the quest for human rights was still exotic. The great English social activist Henry Stephens Salt, who campaigned against the death penalty, co-founded the British Humanitarian League, and propagated vegetarianism, started his trailblazing study entitled *Animals' Rights* with the opening sentence: 'Have the lower animals 'rights'? Undoubtedly—if men have.'[36] In 1892, Salt noted that human ('men's') rights were 'looked upon with suspicion and disfavour by many social reformers', and Salt basically used the term in quotations marks only.[37]

---

[34]Cf. CJEU, *Liga van Moskeeën en islamitische Organisaties Provincie Antwerpen et al v. Vlaams Gewest*, Grand Chamber Judgment of 29 May 2018, Case C-426/16; Peters, 'Religious Slaughter' 2019.

[35]Kymlicka/Donaldson, 'Animal Rights' 2014, 127.

[36]Salt, *Animals' Rights* 2013, 1.

[37]Ibid.

In the decades to follow, the quotation marks around the 'rights of men' disappeared. After 1948, the terms of the UDHR even guided protection for animals. For example, the Preamble of the UDHR proclaims 'the advent of a world in which human beings shall enjoy *freedom of speech and belief and freedom from fear and want*'.[38] These famous four freedoms inspired the so-called 'five freedoms for farm animals' of the 1965 Brambell report[39]: Freedom from hunger and thirst, freedom from discomfort, freedom from injury, pain or disease, freedom to express normal behaviour, and freedom from fear and distress. These 'freedoms' could be creatively understood as legal rights, and could be complemented by further, more fundamental rights such as the right to life and liberty.

Along this line, at the occasion of the 50th anniversary of the UDHR in 1978, a 'Universal Declaration on Animal Rights' (UDAR) was elaborated by an NGO coalition in deliberate alignment with the UDHR.[40] This Animal Rights Declaration was revised in 1989 and again in 2018.[41] The 1978 version of its Art. 1 was modelled on Art. 1 UDHR and runs: 'All animals are born with an equal claim on life and the same rights to existence.'[42] The UDAR was formally proclaimed in 1978 in the UNESCO premises in Paris (albeit not by UNESCO).[43] Although this ceremony attracted a lot of public and media attention, the declaration did not in the long run result in palpable practical effects. Neither have the academic (both philosophical and legal) debates on animal rights—ongoing since the 1960s—led to any serious international codification. It is now time to tackle international animal rights not only at the NGO-level but among governments.

The classic argument in favour of moral duties towards animals has been that prohibiting cruelty on animals suppresses callousness in men.[44] This consideration has traditionally motivated animal welfare laws. It could and should also motivate more ambitious animal rights codifications. Along that line, the preamble of the

---

[38]These 'freedoms' draw on US President Franklin D. Roosevelt's 'four freedoms' proclaimed in the State of the Union address on January 6, 1941. The President proposed four fundamental freedoms that people 'everywhere in the world' ought to enjoy.

[39]Rogers Brambell (Chairman), *Report of the Technical Committee to Enquire into the Welfare of Animals Kept under Intensive Livestock Husbandry Systems* (London: HMSO 1965), 13.

[40]Neumann, 'Universal Declaration' 2012.

[41]LFDA (Fondation Droit Animal, Éthique et Sciences), 'Déclaration des droits de l'animal' (2018), available at: http://www.fondation-droit-animal.org/la-fondation/declaration-des-droits-de-lanimal/.

[42]Art. 1 of the significantly revised version of 1989 was: 'All animals have equal rights to exist within the context of biological equilibrium. This equality of rights does not overshadow the diversity of species and of individuals.' Art. 1 of the 2018-version (n. 41) is available in French only: 'Le milieu naturel des animaux à l'état de liberté doit être préservé afin que les animaux puissent y vivre et évoluer conformément à leurs besoins et que la survie des espèces ne soit pas compromise.'

[43]UNESCO was the site probably because the author of an initial text of 1972, Georges Heuse, was a member of the UNESCO Secretariat (Neumann, 'Universal Declaration' 2012, 95).

[44]Kant, *Metaphysics of Morals* 2015, § 17 'Doctrine of virtue', 192-193.

UDAR of 1978 had stated 'that the respect of humans for animals is inseparable from the respect of man for another man'. The intuition that there is a 'link' between animal abuse and violence against humans has been frequently investigated in sociological and criminological research—with contradictory results.[45] The assertion of such a 'link' is however a double-edged sword. On the one hand, the ruthless criminal prosecution of dog-and-cat-abusers, exactly exploiting this intuition,[46] risks to criminalise low income people, to perpetuate racial stereotypes and actually deflects concern from animals.[47] On the other hand, it has also been shown that the belief in a rigid human—animal divide seems to condone the dehumanisation of humans.[48] The acknowledged need to combat such dehumanisation is an argument in favour of dismantling the legal species hierarchy.[49] The incident with US President Trump at the US-Mexican border demonstrates its relevance.

And because the probably most powerful symbol against such a hierarchy would be the institution of animal rights, the legalisation of some relevant rights for some non-human animals (notably the right to life, liberty, and freedom from torture), should be seriously considered—as a complement to the UDHR. This anthropocentric rationale for animal rights might appeal to different audiences than animal-centred arguments do, and could contribute to building an 'overlapping consensus'[50] on animal rights. We should not wait until a human, ecological, or health-related[51] catastrophe comparable to the horrors which motivated the adoption of the UDHR occurs. An international animal rights codification would not only offer a window of opportunity for mitigating animal suffering but would additionally create positive synergies with the UDHR towards fulfilling its core mission which is to prevent the commission of 'barbarous acts which [outrage] the conscience of mankind', as the Declaration's preamble says.

[45]See critically and with numerous references, ultimately concluding that the intuitive link is not supported by sociological and criminological evidence: Marceau, *Beyond Cages* 2019, chapter 6 (193-250).

[46]This is the current situation in the United States where numerous 'link' programmes seek 'to stop violence against people and animals', as the US 'National link coalition' says. See http://nationallinkcoalition.org/.

[47]Marceau (n. 45), at 274, pointing out that this prosecution practice also entrenches a hierarchy among high and low animals, and demeans low-status humans based on their treatment of high-status animals.

[48]Costello/Hodson, 'Explaining Dehumanization Among Children' 2014. See Kymlicka, 'Human Rights without Human Supremacism' 2018, 763-792, with further references.

[49]As feminism has taught us, the problem are neither the real differences between species nor necessarily their different treatment, but the moral, social, and political hierarchy manifest in differential treatment.

[50]Rawls, 'Overlapping Consensus' 1987.

[51]Yamada/Kahn/Kaplan/Monath/Woodall/Conti, *Confronting Emerging Zoonoses* 2014.

References

Albertson Fineman, M. (2008). The vulnerable subject: Anchoring equality in the human condition. *Yale Journal of Law and Feminism, 20*, 1–23.

Bain, P., Vaes, J., & Leyens, J.-P. (Eds.). (2014). *Humanness and dehumanization.* Abingdon: Routledge.

Baldwin, R., Cave, M., & Lodge, M. (2012). *Understanding regulation: Theory, strategy, and practice* (2nd ed.). New York: Oxford University Press.

Coetzee, J. M. (1999). *The lives of animals.* Princeton: Princeton University Press.

Costello, K., & Hodson, G. (2014). Explaining dehumanization among children: The interspecies model of prejudice. *British Journal of Social Psychology, 53*, 175–197.

Cruft, R., Liao, S. M., & Renzo, M. (2015). The philosophical foundations of human rights: An overview. In R. Cruft, S. M. Liao, & M. Renzo (Eds.), *Philosophical foundations of human rights* (pp. 1–41). Oxford: Oxford University Press.

Dembour, M.-B. (2017). Critiques. In D. Moeckli, S. Shah, & S. Sivakumaran (Eds.), *International human rights law* (3rd ed., pp. 41–59). Oxford: Oxford University Press.

Fischer-Lescano, A. (2018). Natur als Rechtsperson. *Zeitschrift für Umweltrecht, 29*, 205–217.

Kant, I. (2015). *The metaphysics of morals* (19th ed.). Cambridge: Cambridge University Press (German original 1797).

Kurki, V. A. J. (2017). Why things can hold rights: Reconceptualizing the legal person. In V. A. J. Kurki & T. Pietrzykowski (Eds.), *Legal personhood: Animals, artificial intelligence and the unborn* (pp. 69–89). Cham: Springer.

Kymlicka, W. (2018). Human rights without human supremacism. *Canadian Journal of Philosophy, 48*, 763–792.

Kymlicka, W., & Donaldson, S. (2014). Animal rights, multiculturalism, and the left. *Journal of Social Philosophy, 45*, 116–135.

Malanczuk, P. (1997). *Akehurst's modern introduction to international law* (7th rev. ed.). London: Routledge.

Manner, G. (1952). The object theory of the individual in international law. *American Journal of International Law, 46*, 428–449.

Marceau, J. (2019). *Beyond cages: Animal law and criminal punishment.* Cambridge: Cambridge University Press.

Nedelsky, J. (2011). *Law's relations: A relational theory of self, autonomy, and law.* Oxford: Oxford University Press.

Neumann, J.-M. (2012). The universal declaration of animal rights or the creation of a new equilibrium between species. *Animal Law, 19*, 63–109.

Nussbaum, M. (2006). *Frontiers of justice: Disability, nationality, species membership.* Cambridge: Harvard University Press.

Peters, A. (2016). Liberté, Égalité, Animalité: Human-animal comparisons in law. *Transnational Environmental Law, 5*, 25–53.

Peters, A. (2019). Religious slaughter and animal welfare revisited. *Canadian Journal of Comparative and Contemporary Law, 5*, 269–297.

Pinker, S. (2011). *The better angels of our nature: Why violence has declined.* New York: Viking.

Radbruch, G. (1932). Rechtsphilosophie. In G. Radbruch (Ed.), *Gesamtausgabe Band 2: Rechtsphilosophie II* (3rd ed., pp. 206–450). Heidelberg: Müller. 1987 (1993).

Rawls, J. (1987). The idea of an overlapping consensus. *Oxford Journal of Legal Studies, 7*, 1–25.

Raz, J. (2011). Human Rights without foundations. In S. Besson & J. Tasioulas (Eds.), *The philosophy of international law* (pp. 321–337). Oxford: Oxford University Press.

Salt, H. (2013). *Animals' rights: Considered in relation to social progress.* Miami: HardPress Publishing (Originally published New York: Macmillan & Co 1892; repr. of the revised ed. 1922).

Söhner, S. (2016). Habeas Corpus-Beschwerden zugunsten von Menschenaffen. *Rechtswissenschaft, 7*, 556–564.

Stucki, S. (2016). *Grundrechte für Tiere*. Baden-Baden: Nomos.
Sueur, C. (2016). La fuite de la recherche biomédicale sur les primates en Chine: quelles implications éthiques?, *Droit animal éthique & sciences: Revue trimestrielle de la Fondation LFDA* 90.
Triepel, H. (1899). *Völkerrecht und Landesrecht*. Leipzig: Hirschfeldt.
Tushnet, M. (1984). An essay on rights. *Texas Law Review, 62*, 1363–1403.
Yamada, A., Kahn, L. H., Kaplan, B., Monath, T. P., Woodall, J., & Conti, L. (Eds.). (2014). *Confronting emerging zoonoses – The one health paradigm*. Tokyo: Springer.

Anne Peters   is Director at the Max Planck Institute for Comparative Public Law and International Law in Heidelberg, Professor at the Universities of Heidelberg, Freie Universität Berlin and Basel, and a William W. Cook Global Law Professor at the University of Michigan. She has been a member of the European Commission for Democracy through Law (Venice Commission) in respect of Germany (2011–2015) and served as the president of the European Society of International Law (2010–2012) and the German Association of International Law (DGIR, since 2019). Her current research interests relate to public international law including its history, global animal law, global governance and global constitutionalism, and the status of humans in international law.

Chapter 11
(Certified) Humane Violence? Animal Production, the Ambivalence of Humanizing the Inhumane, and What International Humanitarian Law Has to Do with It

Saskia Stucki

Abstract The chapter draws a comparison with the self-certifying of production methods as 'humane' or animal-friendly in the labelling of animal products—that is, according to companies' own self-imposed codes of conduct. It likens the idea of humanizing animal slaughter, factory farms, and other forms of production to the notion of humanizing warfare. Like international humanitarian law (IHL), animal welfare law is marked by the tension inherent in its attempt to humanize innately inhumane practices. Given these parallels, the analysis of animal welfare law might benefit from existing insights into the potential and limits of IHL. Both areas of law endorse a principle of 'humanity' while arguably facilitating and legitimizing the use of violence, and might thereby ultimately perpetuate the suffering of living beings. The implicit justification of violence percolating from the IHL-like animal 'protection' laws could only be outweighed by complementing this body of law with a *ius contra bellum* for animals.

Revised version of the original published article "(Certified) Humane Violence? Animal Welfare Labels, the Ambivalence of Humanizing the Inhumane, and What International Humanitarian Law Has to Do with It" by Saskia Stucki, American Journal of International Law Unbound, Volume 111, 2017, pp. 277–281. The original article was published as an Open Access article, distributed under the terms of the Creative Commons Attribution licence (http://creativecommons.org/licenses/by/4.0/).

S. Stucki (✉)
Max Planck Institute for Comparative Public Law and International Law, Heidelberg, Germany

Harvard Law School, Cambridge, MA, USA
e-mail: stucki@mpil.de

© The Author(s) 2020   121
A. Peters (ed.), *Studies in Global Animal Law*, Beiträge zum ausländischen öffentlichen Recht und Völkerrecht 290,
https://doi.org/10.1007/978-3-662-60756-5_11

1   Introduction: The Industrialization and Humanization of Animal Production

The contemporary human-animal relationship is highly ambivalent. It is characterized by both the exacerbating exploitative use of animals and a progressing moral concern for the life, integrity and welfare of animals. With regard to the agricultural use of animals (which is the quantitatively most significant area of animal use and accounts for more than sixty billion land animals slaughtered globally each year), these two poles stand in particular contrast. On the one hand, agriculture has been increasingly industrialized and intensified over the course of the Twentieth Century. The modern system of industrialized animal production (or the 'animal-industrial complex')[1] is marked by a high degree of rationalization, automatization, efficiency, mass production and profitability, and has turned animals into mere production units—biomachines that convert feed into meat, milk and eggs. On the other hand, the transformation of agriculture to industrialized animal production has raised grave ethical concerns, and societal discomfort at the systemic disregard for the welfare of farmed animals has grown. Most people cringe at the sight of footage showing the horrifying conditions prevailing in factory farms and slaughterhouses, and the vast majority of society subscribes to the basic moral principle that inflicting unnecessary pain and suffering on animals is wrong (a dictum also underlying the nearly universal prohibition of animal cruelty and which is so ingrained it could be considered a 'rule of civilization').[2]

While the growing moral concern for animals has not stopped or reversed the process of industrialization of animal production, the juxtaposition of these two antithetical forces generates a strong dialectical tension. One way of reconciling and harmonizing these two conflicting impetuses is the idea of 'humane production'—i.e. the idea of humanizing animal production. This idea informs, for one, mandatory animal welfare laws which set minimum standards to be respected in agricultural practices. The idea of humanizing animal production is embodied even more clearly in voluntary animal welfare (or humane) labels that react to consumers' demands for higher welfare standards beyond that which is minimally required by law.

The latter phenomenon is the starting point of this enquiry.[3] Critics typically view humane labels as instances of 'humane-washing'. While this critique is important, I believe it falls short. As will be shown, the contradiction and limitations inherent to humane labelling merely epitomize a deeper ambivalence that characterizes animal

---

[1] See generally Noske, *Humans and Other Animals* 1989, 22; Twine, 'Addressing the Animal-Industrial Complex' 2013.

[2] As noted by the dissent in a Canadian appeal decision regarding an elephant in a city-run zoo. See Court of Appeal of Alberta, *Reece v. Edmonton (City)*, Judgment of 4 August 2011, 2011 ABCA 238, para. 56; see also Hurnik/Lehman, 'Unnecessary Suffering' 1982, 131-132.

[3] This chapter does not deal with the issue of the compatibility of animal welfare labels with international trade law. For this, see, e.g., Kelch, 'WTO Tuna Labeling Decision' 2012; on animal welfare and the General Agreement on Tariffs and Trade, Sykes, 'Sealing Animal Welfare' 2014.

welfare law in general: the aporia of humanizing an innately inhumane institution. To my knowledge, there is only one other area of law that is confronted with such a similar ambivalence: international humanitarian law. Learning from the laws of war, it will be argued, may offer valuable new insights for a better understanding and the advancement of animal welfare law.

2   Humane Labelling and Humane-Washing

The term 'animal welfare labelling' or 'humane labelling' covers a wide range of government or private animal welfare labels (e.g. 'Certified Humane', 'Animal Welfare Approved', 'Free Range') on animal-based food products that signal to consumers that such products were produced in compliance with high welfare standards and that the animals involved in the production process were treated humanely.[4] As paradigmatically expressed by the Animal Welfare Institute (an advocate for humane farming practices), the underlying idea of 'humane production' is that each phase of a farmed animal's life (breeding, raising, transport and slaughter) 'offers the opportunity for cruelty or compassion', and that for 'each aspect of industrial production, alternative methods that are both humane and economical are possible.'[5] The overall goal, as for example stated by Humane Farm Animal Care which administers the 'Certified Humane Raised and Handled' label, is 'to [improve] the lives of farm animals in food production from birth through slaughter' and to establish 'kinder and more responsible farm animal practices.'[6] Humane labels also cater to a growing niche-market of ethical consumers willing to pay higher prices for animal friendly products, and are thus believed to be a win-win-win situation for producers, consumers and animals.[7]

While the goal of improving the lives of farmed animals is laudable and it seems trivially true that anything is better than nothing, critics contend that humane labelling at best entails marginal rather than substantial improvements in farmed animal welfare. Many humane labels are notoriously vague, unregulated and unenforced, with no meaningful content or oversight and self-imposed welfare standards that often do not (significantly) go beyond reiterations of the legally

---

[4]For an overview of animal welfare labelling in the EU, see Passantino/Conte/Russo, 'Animal Welfare Labelling 2008, 396-399; Veissier/Butterworth/Bock/Roe, 'European Approaches' 2008, 284 et seq.; for the US context, see Wiseman, 'Localism, Labels, and Animal Welfare' 2018, 75 et seq.; Leslie/Sunstein, 'Animal Rights Without Controversy' 2007, 125 et seq.

[5]Animal Welfare Institute, 'Farm Animals', available at: https://awionline.org/content/farm-animals.

[6]Humane Farm Animal Care, 'Our Mission', available at: https://certifiedhumane.org.

[7]On the 'marketization' of farm animal welfare, see generally Buller/Roe, 'Modifying and Commodifying' 2014, (noting, *inter alia*, the performativity of 'doing animal welfare', which has become 'a broad array of technics, practices and materialities to meet reasoning present in the "market", rather than in the sole interest of improving animal welfare.' ibid. 142).

required minimum or reflections of standard agricultural practices.[8] According to critics, humane labels are therefore mostly misleading and primarily amount to a marketing strategy, or as Marc Bekoff puts it, 'feel-good scams' that enable consumers to buy a clean conscience.[9]

This line of criticism is best captured by the term 'humane-washing'.[10] Like greenwashing, humane-washing is a type of whitewashing, which is a metaphor for communications that gloss over or obscure unpleasant, negatively connoted facts. Based on common definitions of greenwashing,[11] humane-washing can be defined as the dissemination of false or deceptive information by companies so as to promote the perception that its products are animal-friendly, or as 'symbolic information emanating from within an organization without substantive actions', measurable as the discrepancy between *saying* ('humane talk') and *being* humane ('humane walk').[12] This, then, is the core of the humane-washing critique: that what is presented by humane labels as humane is in fact not humane.

3   The Inherent Contradiction and Limits of Humanizing Animal Production

While I principally agree with the criticism of humane-washing, I believe that it does not go quite far enough. The issue is not just that what is presented as humane is in fact not humane, but that it cannot be. Put differently, it is not just that humane labels promise something which they factually do not deliver—it is that they envisage something that is actually impossible. As will be shown in this section, the problem thus runs deeper than humane-washing. The contradiction of calling inhumane

[8]For example, in a report titled 'Humanewashed', the Animal Welfare Institute concluded that the animal welfare programs certified by the USDA Process Verified Program as 'humane' were not materially different from conventional production methods. See Rachel Mathews, 'Humanewashed: USDA Process Verified Program Misleads Consumers About Animal Welfare Marketing Claims', March 2012, available at: https://awionline.org/sites/default/files/uploads/documents/fa-humanewashedreportonusdapvp.pdf.

[9]Marc Bekoff, 'Stairways to Heaven' 2016; see also Wrenn, 'Abolitionist Animal Rights' 2012, 443-445; on the 'free-range' egg label, see Parker/Brunswick/Kotey, 'The Happy Hen' 2013, (concluding that 'free-range' egg labels 'are generally either misleading or deceptive and that the notion of "free-range" has been industrialized and watered down so much as not to meet significant animal welfare, environmental, and public health concerns', ibid. 182).

[10]See, e.g., Bekoff/Pierce, *Animals' Agenda* 2017, 50 et seq.; or 'hogwashing', which LaVeck describes as 'the practice of generating the public appearance of having compassion for animals while continuing to kill millions of them for profit.' See LaVeck, 'Compassion for Sale?' 2006.

[11]See, e.g., Becker-Olsen/Potucek, 'Greenwashing' 2013, 1318.

[12]Walker/Wan, 'Harm of Symbolic Actions' 2012, 231.

farming practices 'humane' is not merely an intentionally deceptive marketing strategy,[13] but lies in the very nature of the project of humanizing animal production.

In absolute terms, animal production is inherently inhumane and can, *eo ipso*, not be (fully) humanized. According to the dictionary, to 'humanize' is 'to make humane', which means 'marked by compassion, sympathy, or consideration' for animals.[14] By contrast, even 'humane production' methods regularly involve severe forms of violence against animals, such as confinement, tail docking, dehorning, castration and debeaking without anaesthesia, forced impregnation of milk cows and the separation from their calves, and the mass maceration or gassing of male chicks.[15] Minimally, every kind of (economically sound) animal production will inevitably culminate in the ultimate act of violence: the involuntary and premature death of an animal, e.g. by the cutting of major blood vessels. The US Humane Methods of Slaughter Act gives a glimpse at what 'humane slaughter' looks like: it means to be 'rendered insensible to pain by a single blow or gunshot or an electrical, chemical or other means that is rapid and effective, before being shackled, hoisted, thrown, cast, or cut' (the reality is even grimmer, as many animals are slaughtered in a state of consciousness due to improper stunning).[16] These routine acts of violence can hardly be considered 'humane' in any real sense of the word, all the more considering that they are also quite unnecessary.[17] The only way 'humane' can be negotiated with regard to animal production is in relative terms. In this sense,

---

[13]This discrepancy naturally produces the kind of doublespeak that also resonates in the criticism of humane-washing. For example, in an early critique of animal welfare labels, LaVeck traces how industry stakeholders (in collaboration with professionalized animal organizations) have appropriated, redefined, eroded and commodified the key language of the animal advocacy movement – compassion and humaneness – and channelled it into a new 'happy meat' discourse. What has emerged is a 'brave new world' in which 'a mechanized system designed to move animals quickly and efficiently, to take their lives, to drain their blood, and to cut them into pieces on a scale never before imagined, is proudly described as a "stairway to heaven" by a slaughterhouse designer.' See LaVeck, 'Compassion for Sale?' 2000, (referring to Temple Grandin, who calls her humane slaughterhouse ramp and restraining systems 'stairways to heaven').

[14]See Merriam Webster Dictionary, 'humanize' and 'humane', available at: https://www.merriam-webster.com/dictionary/humanize and https://www.merriam-webster.com/dictionary/humane.

[15]For an overview of standard practices in modern industrialized animal production, see, e.g. Norwood/Lusk, *Compassion, by the Pound* 2011, 94 et seq. *et passim.*

[16]US, Humane Methods of Slaughter Act, 7 U.S.C. § 1901 et seq., section 2 letter (a); on the failures of the Humane Methods of Slaughter Act, see DeCoux, 'Speaking for the Modern Prometheus' 2009, 19 et seq.

[17]A plant-based diet is perfectly healthy at least in modern industrialized societies. See generally Melina/Craig/Levin, 'Position of the Academy of Nutrition' 2016, 1970-1980; as noted by Kymlicka and Donaldson, 'virtually all human violence against animals is unnecessary in the strict sense. Since humans can lead flourishing lives without eating meat, or wearing leather, or visiting caged animals in zoos or circuses, none of the suffering involved in these practices is necessary.' Kymlicka/Donaldson, 'Animal Rights' 2014, 126; on the notion of necessity in the context of animal suffering, see generally Francione, *Introduction to Animal Rights* 2007b, 9 and 55 et seq.

humanization is merely an approximation, meaning 'more humane' compared to pre-existing 'worst-case scenarios' that are even more inhumane.[18] Within the bounds of this relative standard, then, the (slightly) less inhumane becomes 'humane', a (slightly) less horrible life becomes a 'good life', and a (slightly) less miserable animal becomes a 'happy animal', or 'happy meat'.[19]

Now, the same holds true for animal welfare regulation at large. Generally, the law plays an ambivalent role in governing the human-animal relationship, setting the parameters for both the protection and exploitation of animals. More specifically, as I have argued at length elsewhere, contemporary animal welfare law (AWL) is structurally constituted by the ambivalence of humanizing the inherently inhumane and violent institution of animal exploitation.[20] In view of its historical formation, AWL has emerged as a secondary function of and as reaction to the pre-existing practices of animal use and the ethical issues they raise. The normative purpose of AWL is not to do away with this institution as such, but rather to humanize it in relative terms (i.e., make it more humane, or less inhumane) by regulating the modalities of permissible violence against animals and thereby mitigating, to some extent, the suffering caused.[21] In doing so, animal welfare regulation perpetuates a somewhat paradoxical and reactive dynamism of imperfectly humanizing while legalizing, facilitating and reinforcing the very institution that inevitably exerts violence against animals and makes them vulnerable and in need of protection in the first place.

4   Animal Welfare Law and International Humanitarian Law: A Brief Comparison

Viewed in this light, to the extent that AWL may be characterized as a body of law whose quintessential function is to approximatively humanize a profoundly inhumane institution, it bears a striking prima facie resemblance with another, quite unrelated body of law. Namely, there is—to my knowledge—only one other area of

---

[18]See Bekoff/ Pierce, *The Animals' Agenda* 2017, 50-51 (further noting that 'humane' is 'one of the most overused and meaningless [words] in our current vocabulary').

[19]The term 'happy meat' encapsulates the 'belief that it is possible to raise and kill animals in such a way as to remove the ethical problems' associated with the notion of 'animal machines' and factory farming. See Cole, 'From "Animal Machines" to "Happy Meat"' 2011, 84.

[20]See Stucki, *Grundrechte für Tiere* 2016b, 140-149.

[21]This reactive dynamism of mitigation rather than prevention was identified clearly by the Israeli High Court of Justice in a case concerning the production of foie gras. Commenting on the 'problematic language' of the relevant regulation, the Court remarked that the stated 'purpose of the Regulations is to "prevent the geese's suffering." Clearly these regulations do not prevent suffering; at best they minimize, to some extent, the suffering caused.' Supreme Court of Israel (sitting as the High Court of Justice), *Noah v. Attorney General*, Verdict of 11 August 2003, Appeal No. 9232/01, para. 17 (judgment of Justice A. Grunis).

law that operates with such a deeply ambivalent dynamic: international humanitarian law (IHL), that is, the legal regime governing war and other armed conflict. While the idea of comparing AWL and IHL might seem far-fetched, there exist significant parallels between these two legal regimes, which I will briefly highlight in this section. With a view to future research, I believe these parallels merit closer attention, and exploring them will not only be analytically interesting, but may also be helpful for furnishing a more accurate understanding of the nature and limits of AWL.

To start with, IHL, much like AWL, is marked by the inherent contradiction of humanizing an innately inhumane and violent institution: war and other armed conflict. As noted by Theodor Meron, in order to 'genuinely humanize humanitarian law, it would be necessary to put an end to all kinds of armed conflict. But wars have been part of the human condition. . .and regrettably they are likely to remain so.'[22] In recognizing the (unfortunate) reality of war, the purpose of IHL is not the (manifestly utopian) absolute elimination of the calamities of war by prohibiting warfare as such, but its rendering more humane, however imperfectly, by reducing the human suffering caused in its course.[23] However, despite its terminology (which would seem to suggest otherwise), humanitarianism is not the sole objective, but just one among other conflicting purposes of IHL.[24] Notably, IHL is constituted by a dialectical tension between two diametrically opposed forces—*military necessity* versus *humanitarian/human welfare considerations* – and aims at a compromise by 'minimizing human suffering without undermining the effectiveness of military operations.'[25] Similarly, AWL (whose terminology might also misleadingly suggest that its sole legislative purpose is animal welfare, whereas the latter is just one among other conflicting purposes) reflects a compromise between two diametrically opposed factors, notably *economic* and other instrumental *necessities* on the one hand and *humane/animal welfare considerations* on the other hand.[26]

Interestingly, both AWL and IHL further operate with the key principle of 'unnecessary suffering'. In the context of IHL, the basic rule prohibiting the use of means and methods of warfare of a nature to cause superfluous injury or unnecessary suffering intends to lower the cruel effects of weapons on combatants.[27] The International Court of Justice has defined unnecessary suffering as 'harm greater

---

[22]Meron, 'Humanization of Humanitarian Law' 2000, 240.

[23]See also Tomuschat, 'Human Rights' 2010, 16 (noting that IHL 'is designed to ensure a minimal protection even during the most profound catastrophe of human society, namely war' and seeks 'to salvage what realistically can be protected notwithstanding the clash of arms').

[24]On the misleading terminology of international *humanitarian* law, see Wilson, 'Myth of International Humanitarian Law' 2017, 563-579, *passim* (challenging the conventional narrative that equates what was traditionally the 'law of war' or 'law of armed conflict' with the modern term of 'international humanitarian law').

[25]Dinstein, *Conduct of Hostilities* 2016, 8-9.

[26]See Stucki, *Grundrechte für Tiere* 2016b, 141-146.

[27]Protocol Additional to the Geneva Conventions of 12 August 1949, and relating to the Protection of Victims of International Armed Conflicts (Protocol I), 8 June 1977, 1125 UNTS 3, art. 35(2); on

than that unavoidable to achieve legitimate military objectives',[28] that is, suffering that has no military purpose. Similarly, in the context of AWL, the term 'unnecessary suffering' is typically applied to inflictions of wanton and gratuitous suffering that 'goes beyond what is necessary for "appropriate" exploitation' and that serves no legitimate (e.g., economic) purpose.[29]

Ironically, while IHL serves the compelling goal of humanizing armed conflict, it also enhances its acceptability, has a certain affirming and legitimizing effect, and may thus even prolong the residual violence entailed by the institution of war as such.[30] A similar point is made in the context of AWL, which, as critics contend, may effectively serve to legitimize and reinforce exploitative animal use by making it more socially acceptable.[31] (The legitimacy of the institution as such is logically implicit, given that if it were considered illegal and profoundly illegitimate, the law would prohibit rather than simply regulate the institution, as for example in the case of slavery). Overall, IHL, much like AWL, thus embodies the ambivalence of humanizing, while simultaneously facilitating and consolidating, the very institution that is the cause of the violence and suffering it aims to mitigate.

---

the principle of unnecessary suffering, see generally Meyrowitz, 'Principle of Superfluous Injury' 1994.

[28]ICJ, Legality of the Threat or Use of Nuclear Weapons, Advisory Opinion of 8 July 1996, ICJ Reports 1996, 226, para. 78.

[29]Francione, *Animals, Property, and the Law* 2007a, 146; on the notion of 'unnecessary suffering' in applied animal welfare law, ibid., 17 et seq. and 142 et seq.; for example, German courts have held that the killing of around 45 million male chicks annually due to their economic uselessness – a standard practice in the egg industry – constitutes a 'necessary' and thus lawful infliction of harm on animals. This justifying 'necessity' is a purely economic necessity and accrues from the fact that economically viable alternatives to chick culling are not available so far. See, notably, OVG Münster, Judgments of 20 May 2016, Case No. 20 A 488/15 and 20 A 530/15 (the temporary legality of killing male chicks was confirmed by the Federal Administrative Court, Judgments of 13 June 2019, Case No. 3 C 28.16 and 3 C 29.16); on the German 'male chick'-judgments and the underlying notion of 'necessity', see Stucki, 'Die Nutzung kommt vor dem Schutz' 2016a.

[30]See Meron, 'The Humanization of Humanitarian Law' 2000, 241; on the limits and legitimizing effects of IHL in the context of occupation, see generally Gross, *Writing on the Wall* 2017.

[31]See, e.g., Donaldson/Kymlicka, *Zoopolis* 2011, 2 (noting that the marginal improvements achieved through animal welfare reforms at best 'distract attention from the underlying system of animal exploitation, and at worst, they provide citizens with a way to soothe their moral anxieties, providing false reassurance that things are getting better, when in fact they are getting worse. (...) these ameliorist reforms serve to legitimate, rather than contest, the system of animal exploitation'); Francione, *Rain Without Thunder* 1996), 36-37 (stating that animal welfare regulation 'seeks to reform institutions of animal exploitation and make them more "humane" and explicitly reinforces the moral orthodoxy of human hegemony over nonhumans'); Bourke, 'Use and Misuse of "Rights Talk"' 2009, 133 (noting that 'animal welfare legislation is often used not just to protect animals but also to regulate, and indeed facilitate, the ongoing use of animals'); Wrenn, 'Abolitionist Animal Rights' 2012, 446.

5   Outlook: From a 'Jus In Bello' Towards a 'Jus Contra Bellum' for Animals

As this brief comparison indicates, there exist some remarkable parallels between the structure, function and effects of AWL and IHL. Even though these parallelisms will require further exploration and reflection, the comparison between IHL and AWL casts a new light onto AWL and allows for some thought-provoking conclusions. First of all, it suggests that the institutional and systemic practice of violence predominantly shaping the contemporary human-animal relationship could be conceptualized as constituting something akin to a 'war against animals'.[32] Secondly, and accordingly, AWL may be best understood as a kind of 'jus in bello' that governs and regulates violent activities within this 'war' on animals. Thirdly, while AWL—so understood—serves an important, yet contradictory, limited, and even legitimizing humanizing function by alleviating suffering as long as the reality of 'war' factually persists, such a 'jus in bello' can neither justify the 'war' as such nor does it suffice.

Drawing from Aeyal Gross, this calls for a shift from a 'merely factual' to a 'normative' approach,[33] one that does not merely posit the 'war' on animals as an (ugly) fact, but acknowledges its normative dimension. This expanded perspective should address questions not just specifically relating to the appropriate conduct in 'war', but more fundamentally relating to the legitimacy of the 'war' as such. Most crucially, what is needed first and foremost, and what is lacking so far in the case of animals, is for the 'jus in bello' to be complemented by a higher-priority set of norms that work to prevent a state of 'war' in the first place—as it were, a kind of 'jus contra bellum' for animals.

References

Becker-Olsen, K., & Potucek, S. (2013). Greenwashing. In S. O. Idowu et al. (Eds.), *Encyclopedia of corporate social responsibility* (pp. 1318–1323). Heidelberg: Springer.
Bekoff, M. (2016, November 17). Stairways to Heaven, Temples of Doom, and Humane-Washing, *Psychology Today*, available at: https://www.psychologytoday.com/us/blog/animal-emotions/201611/stairways-heaven-temples-doom-and-humane-washing.
Bekoff, M., & Pierce, J. (2017). *The animals' agenda: Freedom, compassion, and coexistence in the human age*. Boston: Beacon Press.

---

[32]This is indeed proposed by Wadiwel, *War against Animals* 2015; Wadiwel further suggests that this 'is a war that operates under the guise of peace, constructed more often than not within the rule of law. This war does not appear to be a war'. Wadiwel, 'War against Animals: Domination, Law and Sovereignty'2009, 285.

[33]Cf. Gross, *The Writing on the Wall* 2017, 3-4 (calling for a distinction between *'jus in occupation'* and *'jus ad occupation'*).

Bourke, D. (2009). The use and misuse of "Rights Talk" by the animal rights movement. In P. Sankoff & S. White (Eds.), *Animal law in Australasia: A new dialogue* (pp. 128–150). Annandale: Federation Press.

Buller, H., & Roe, E. (2014). Modifying and commodifying farm animal welfare: The economisation of layer chickens. *Journal of Rural Studies, 33*, 141–149.

Cole, M. (2011). From "Animal Machines" to "Happy Meat"? Foucault's ideas of disciplinary and pastoral power applied to "Animal-Centred" welfare discourse. *Animals, 1*, 83–101.

DeCoux, E. L. (2009). Speaking for the modern prometheus: The significance of animal suffering to the abolition movement. *Animal Law Review, 16*, 9–64.

Dinstein, Y. (2016). *The conduct of hostilities under the law of international armed conflict* (3rd ed.). Cambridge: Cambridge University Press.

Donaldson, S., & Kymlicka, W. (2011). *Zoopolis: A political theory of animal rights*. Oxford: Oxford University Press.

Francione, G. L. (1996). *Rain without thunder: The ideology of the animal rights movement*. Philadelphia: Temple University Press.

Francione, G. L. (2007a). *Animals, property, and the law*. Philadelphia: Temple University Press reprinted.

Francione, G. L. (2007b). *Introduction to animal rights: Your child or the dog?* Philadelphia: Temple University Press reprinted.

Gross, A. (2017). *The writing on the wall: Rethinking the international law of occupation*. Cambridge: Cambridge University Press.

Hurnik, F., & Lehman, H. (1982). Unnecessary suffering: Definition and evidence. *International Journal for the Study of Animal Problems, 3*, 131–137.

Kelch, T. G. (2012). The WTO Tuna labeling decision and animal law. *Journal of Animal and Natural Resource Law, 8*, 121–138.

Kymlicka, W., & Donaldson, S. (2014). Animal rights, multiculturalism, and the left. *Journal of Social Philosophy, 45*, 116–135.

LaVeck, J. (2006, September). Compassion for sale? Doublethink meets doublefeel as happy meat comes of age, *Satya*, Available at: http://www.satyamag.com/sept06/laveck.html

Leslie, J., & Sunstein, C. R. (2007). Animal rights without controversy. *Law and Contemporary Problems, 70*, 117–138.

Melina, V., Craig, W., & Levin, S. (2016). Position of the academy of nutrition and dietetics: Vegetarian diets. *Journal of the Academy of Nutrition and Dietetics, 116*, 1970–1980.

Meron, T. (2000). The humanization of humanitarian law. *American Journal of International Law, 94*, 239–278.

Meyrowitz, H. (1994). The principle of superfluous injury or unnecessary suffering: From the declaration of St. Petersburg of 1868 to Additional Protocol I of 1977. *International Review of the Red Cross, 34*, 98–122.

Norwood, F. B., & Lusk, J. L. (2011). *Compassion, by the pound: The economics of farm animal welfare*. Oxford: Oxford University Press.

Noske, B. (1989). *Humans and other animals: Beyond the boundaries of anthropology*. London: Pluto Press.

Parker, C., Brunswick, C., & Kotey, J. (2013). The happy hen on your supermarket shelf: What choice does industrial strength free-range represent for consumers? *Bioethical Inquiry, 10*, 165–186.

Passantino, A., Conte, F., & Russo, M. (2008). Animal welfare labelling and the approach of the European Union: An overview on the current situation. *Journal für Verbraucherschutz und Lebensmittelsicherheit, 3*, 396–399.

Stucki, S. (2016a). Die Nutzung kommt vor dem Schutz – und andere Lehren aus der neuen Küken-Rechtsprechung. *Rechtswissenschaft, 7*, 521–541.

Stucki, S. (2016b). *Grundrechte für Tiere: Eine Kritik des geltenden Tierschutzrechts und rechtstheoretische Grundlegung von Tierrechten im Rahmen einer Neupositionierung des Tieres als Rechtssubjekt*. Baden-Baden: Nomos.

Sykes, K. (2014). Sealing animal welfare into the GATT exceptions: The international dimension of animal welfare in WTO disputes. *World Trade Review, 13*, 471–498.

Tomuschat, C. (2010). Human rights and international humanitarian law. *European Journal of International Law, 21*, 15–23.

Twine, R. (2013). Addressing the animal-industrial complex. In R. Corbey & A. Lanjouw (Eds.), *The politics of species: Reshaping our relationships with other animals* (pp. 77–92). Cambridge: Cambridge University Press.

Veissier, I., Butterworth, A., Bock, B., & Roe, E. (2008). European approaches to ensure good animal welfare. *Applied Animal Behaviour Science, 113*, 279–298.

Wadiwel, D. J. (2009). The war against animals: Domination, law and sovereignty. *Griffith Law Review, 18*, 283–297.

Wadiwel, D. J. (2015). *The war against animals*. Leiden: Brill.

Walker, K., & Wan, F. (2012). The harm of symbolic actions and green-washing: Corporate actions and communications on environmental performance and their financial implications. *Journal of Business Ethics, 109*, 227–242.

Wilson, P. (2017). The myth of international humanitarian law. *International Affairs, 93*, 563–579.

Wiseman, S. R. (2018). Localism, labels, and animal welfare. *Northwestern Journal of Law & Social Policy, 13*, 66–83.

Wrenn, C. L. (2012). Abolitionist animal rights: Critical comparisons and challenges within the animal rights movement. *Interface, 4*, 438–458.

Saskia Stucki   is a visiting scholar at the Harvard Animal Law and Policy Program (2018–2019), Senior Research Fellow at the Max Planck Institute for Comparative Public Law and International Law (since 2016), and recipient of an Advanced Postdoc. Mobility scholarship from the Swiss National Science Foundation SNSF (2018–2020). Her research interests cover animal law and animal rights, human rights philosophy, and legal theory. She is currently working on her SNSF-funded postdoctoral research project "Trilogy on a Legal Theory of Animal Rights".

Part IV
New Protective Legal Strategies

Chapter 12
Trophy Hunting, the Race to the Bottom, and the Law of Jurisdiction

Charlotte E. Blattner

Abstract Cross-border trade, industry outsourcing, and animal migration are increasingly challenging states that want to take their commitment to protecting animals seriously. When multinationals threaten to outsource, even the most powerful states succumb to economic pressure and give corporations what they so avidly desire: laissez-faire. Some argue this is an inevitable consequence of globalization; others say it prompts us to question whether animal law is not better off being regulated by international law. This chapter takes a third path. Instead of proposing that nations seek agreement on low and mostly ineffective animal welfare standards, it posits extraterritorial jurisdiction as a promising avenue for animal law, and takes trophy hunting as its example to illustrate the many jurisdictional options for states to overcome regulatory gaps in animal law and make animal issues more visible on the international plane.

1 Introduction

Cross-border trade, industry outsourcing, and animal migration arc increasing challenges for states that want to take their commitment to protect animals seriously. When multinationals threaten to outsource, even the most powerful states succumb to economic pressure and give corporations what they so avidly desire: *laissez-faire*. Some argue this is the inevitable consequence of globalization, others say it should

Revised version of the original published article "Can Extraterritorial Jurisdiction Help Overcome Regulatory Gaps of Animal Law? Insights from Trophy Hunting" by Charlotte Blattner, American Journal of International Law Unbound, Volume 111, 2017, pp. 419–424. The original article was published as an Open Access article, distributed under the terms of the Creative Commons Attribution licence (http://creativecommons.org/licenses/by/4.0/).

C. E. Blattner (✉)
Harvard University, Law School, Cambridge, MA, USA
e-mail: cblattner@law.harvard.edu

A. Peters (ed.), *Studies in Global Animal Law*, Beiträge zum ausländischen öffentlichen Recht und Völkerrecht 290,
https://doi.org/10.1007/978-3-662-60756-5_12

make us question whether animal law ought to be territorially bound and local, and emphasize the need for an international treaty in animal law. This chapter takes a third path. Instead of proposing that nations seek agreement on low and mostly ineffective animal welfare standards, it argues that extraterritorial jurisdiction is the more promising avenue for animal law, and takes trophy hunting as its example to illustrate the many jurisdictional options for helping animal law overcome regulatory gaps and making animal issues more visible on the international plane.

2   Regulating Trophy Hunting in an Era of Globalization: A Lost Cause?

In 2015, the world was outraged to hear that Cecil, a black-maned lion, was shot and killed by an American game hunter in Zimbabwe. Cecil was a resident of the Hwange National Park, where he was a star attraction for many visitors and part of a long-term national study on lion movement. Cecil was lured out of the park by carcasses tied to a car, and then shot with a bow and arrow by Walter Palmer, a US citizen who paid 50,000 USD to kill Cecil and claim his remains. Severely wounded, Cecil ran from the hunters for more than 40 h before they fired the fatal shot. When the public learned of these events, Palmer faced what some journalists described as 'a global storm of internet indignation,' and 'an online witch-hunt'.[1]

Though Cecil's killing got abundant media coverage and sparked public outrage, many other such killings for trophies go unremarked. In trophy hunting (so-called sport or recreational hunting), animals are killed for their head, horns, paws, or skin.[2] Typically, hunters target the rarest and biggest animals, or those who are hardest to chase and shoot. Trophy hunting is practiced in many states, but has been subject to increased public scrutiny in the US due to its high imports. According to the Humane Society International, the US imported 1.26 million wildlife trophies between 2005 and 2014.[3] Most trophies originated in Canada and South Africa; a smaller number came from Argentina, Botswana, Mexico, Namibia, New Zealand, Tanzania, and Zambia. Trophy hunters are known to pay large sums to kill exotic animals and take possession of their dead bodies. For an African lion, trophy hunters pay between 13,500 and 49,000 USD and for an African elephant, between 11,000 and 70,000 USD. Among the animals hunted and imported into the US, 32,230 were members of

---

[1]Capecchi/Rogers, 'Killer of Cecil the Lion' 2015.

[2]The term trophy hunting does not indicate whether it is legal at the place where the animal is killed; poaching, in contrast, clearly denotes illegal wildlife killing.

[3]Humane Society International (HSI), 'Trophy Hunting by the Numbers: The United States' Role in Global Trophy Hunting', 2016, available at: http://www.hsi.org/assets/pdfs/report_trophy_hunt ing_by_the.pdf.

the African 'Big Five': 5600 African lions, 4600 African elephants, 4500 African leopards, 330 Southern white rhinos, and 17,200 African buffaloes.[4]

Although the US prohibits the importation of (at least some) trophies under the Endangered Species Act (ESA),[5] illegal trade of trophies continues unabated. One reason for this, as is claimed on a recurring basis, could be the lack of enforcement of the ESA, or of the Convention on International Trade in Endangered Species (CITES)[6] upon which the ESA is based. Another reason could be that trophy hunting is still legal in more than twenty African countries,[7] so regulating the importation of trophies does little to stop the ongoing endangerment of or threats to wild species. Arguably, the laws of trophy-importing states would be much more effective if they were not to apply at such a late point in time, namely when the animal is dead already. These states would ideally regulate the state of facts earlier, by governing acts of planning, hunting, shooting, and preparing an animal for exportation. Given the fact that—if we take Cecil's case—the lion was killed on Zimbabwean territory, however, prescriptive jurisdiction over trophy hunting *prima facie* seems to lie with Zimbabwe. Any effort on part of importing states to chime in on the Cecil case before the animal crosses the border therefore would seem to violate Zimbabwe's sovereign jurisdiction.

3   Are Treaties the Solution?

Most states recognize that we live in a highly intertwined world, where daily activities across borders easily give rise to state interests reaching beyond domestic territory. A mediated view might therefore suggest that multiple states have a legitimate interest in the case—Zimbabwe as Cecil's home state and the place where his killing took place, and the US as the perpetrator's home state , and urge them to come to an agreement. Proponents of this mediated view might also suggest that the parties must seek to work towards an international treaty to prohibit hunters from killing animals that belong to endangered or threatened species. Such a treaty would ensure that all states' views, preferences, and interests were taken into account, and it would be carried by their willingness to cooperate. Treaty making seems to offer the quickest way to resolve conflict in a manner acceptable to all parties over the long-term, and which is hence likely to be enforced by them.

---

[4]Ibid., 1.

[5]US, Endangered Species Act, 28 December 1973, 16 U.S.C. § 1531.

[6]Convention on International Trade in Endangered Species of Wild Fauna and Flora, 3 March 1973, 993 UNTS 243. *See* on enforcement issues of the CITES: McOmber, 'Problems in Enforcement of the Convention on International Trade in Endangered Species' 2002, 674-701. Ferraro et al. show that listing a species under the ESA is, on average, detrimental to species recovery if not combined with substantial government funds: Ferraro/McIntosh/Ospina, 'The Effectiveness of the US Endangered Species Act' 2007, 245-261.

[7]Onishi, 'Outcry for Cecil the Lion Could Undercut Conservation Efforts' 2015.

But how feasible is this proposal? The difficulty of coming to a broad agreement is easily underestimated, and failure to reach agreement is the rule, rather than the exception. Even in the specific and narrow context of protecting endangered species, states profoundly disagree over the optimal regulatory measures needed to thwart trophy hunting. How can this be explained? In a seminal article on antitrust law, Andrew Guzman used an economic analysis to determine the probability states would conclude an international treaty on jurisdictional matters. He hypothesized that economic incentives are states' primary motive for seeking or rejecting a treaty, and argued that finding common ground for a treaty will be difficult, if not impossible, when consumers and producers are unevenly distributed among states.[8] Let us assume state A is a majority world country,[9] strongly influenced by investors, and state B is a minority world country, presumably investment-exporting and, therefore, more consumer-oriented. According to Guzman, the optimal policy for state A is to have no policy, since welfare losses are borne by consumers abroad. The optimal policy for state B, however, is to regulate at a level that increases efficiency gains for consumers.[10]

Guzman's probability analysis can neatly be extrapolated to animal law, because economic considerations play such an important role in its policy-making, and because a large portion of the world's animal products is produced in the majority world. Let us again hypothesize that state A is investment-driven while state B is more consumer-oriented. For state A, the optimal solution is for animal production to be unregulated, so it will tend to under-regulate. For state B, the optimal solution is regulation that better satisfies consumer preferences, so it will tend to overregulate. Both states are biased to disproportionly protect either producers or consumers. Based on these disparate preferences, the likelihood that these states agree on a set of norms that allocate jurisdictional competence among them is extremely low. Moreover, states in Zimbabwe's position are unlikely to prohibit practices that generate considerable income revenue for them. These considerations show that treaties, designed to determine the jurisdictional parameters of animal law, are a less feasible policy option than they might initially appear.

Even if feasible, concluding an international treaty might not be desirable in the first place. A treaty may frustrate the very reason for which its conclusion is sought,

---

[8]Guzman, 'Is International Antitrust Possible?' 1998, 1501-1548.

[9]In international law, we typically speak of 'developing states' or the 'Third World' to denote countries in juxtaposition to 'developed countries.' These terms imply that development is a standardized and linear process, and that certain states have finished developing while others are still striving to develop. Because states evolve differently, and because their different strengths and challenges should be acknowledged, these terms seem both inaccurate and inappropriate. Scholars are increasingly using the terms 'majority world' and 'minority world' instead. The term 'majority world' highlights the fact that most of the world's population live in regions previously identified as 'developing.' The term 'minority world' refers to countries traditionally identified as 'developed,' in which a minority of the world's population resides. See e.g., Punch, 'Exploring Children's Agency Across Majority and Minority World Contexts' 2016, 183-196.

[10]Guzman concedes that this is the simplest analytical model, yet it allows drawing the best inferences: Guzman, 'Is International Antitrust Possible?' 1998, 1514-1515.

by boiling animal laws to the lowest common denominator and by driving a wedge between different cultures and societies concerning the question of what the 'optimal treatment' of animals is. Also, the risks entered by waiting for an international agreement to form—risks that are born by the animals who are directly and indirectly affected, by local communities that rely on these animals, and by ecosystems in which animals play a key role—, make deferring the issue a poor option.[11] One could argue that the downsides of waiting for an agreement will easily be outbalanced by the benefits of coming to an agreement, but this view greatly underestimates the transaction risks. Since trophy money typically moves to the state that offers hunters the cheapest prices at the lowest level of regulation (hence, to state A), state B's efforts to protect animals will always be undermined.[12] The time allocated to finding an agreement is thus likely time granted to a competition towards laxity, from which animals will suffer most.[13]

4   The Promises of Extraterritorial Jurisdiction

This is where the benefits of extraterritorial jurisdiction come into play. Extraterritorial jurisdiction, for the purposes of the present inquiry, refers to a state's authority to prescribe law over persons, property, or events on foreign territory.[14] Given the diverging views on and within animal law, there is a justified concern that extraterritorial jurisdiction might only exacerbate existing tensions. While these risks can never fully be excluded, judging extraterritorial jurisdiction solely on this basis fails to do justice to the concept and its promises. A noteworthy promise of extraterritorial jurisdiction is that the various forms of overlapping and concurring laws will create a dense jurisdictional net across the globe. This promise is famously defended by Schiff Berman and yields two important benefits.[15] First, the *prima facie* permissi-

---

[11]*See* on the effects of trophy hunting on lion communities: Packer/Brink/Kissui/Maliti/Kushnir/ Caro, 'Effects of Trophy Hunting on Lion and Leopard Populations in Tanzania' 2011, 142-153. *See* on the role of lions in ecosystems: Estes/Crooks/Holt, 'Ecological Role of Predators' (2013), 229-249.

[12]Guzman, 'Is International Antitrust Possible?' 1998, 1523.

[13]I briefly answer the question of whether races to the bottom exist or prevail in the many regulatory areas of animal law in sec. 5.

[14]In other words, we are here dealing with prescriptive jurisdiction. The two other forms of jurisdiction—adjudicative and enforcement jurisdiction—are more problematic under international law: Meng, *Extraterritoriale Jurisdiktion im öffentlichen Wirtschaftsrecht* 1994, 4.

[15]Schiff Berman argues that 'we might deliberately seek to create or preserve spaces for productive interaction among multiple, overlapping legal systems by developing procedural mechanisms, institutions, and practices that aim to manage, without eliminating, the legal pluralism we see around us.' (Schiff Berman, *Global Legal Pluralism: A Jurisprudence of Law Beyond Borders* 2012, 457).

bility of multiple jurisdictional assertions that overlap and concur decreases the likelihood of regulatory gaps in animal law: 'Let both States assert jurisdiction.'[16] Second, extraterritorial jurisdiction creates opportunity for political deliberation and nuanced negotiation, for adapting sweeping or insufficient laws, and leaves space for creative innovation and competition. The legal pluralism that emerges from extraterritorial jurisdiction makes apparent its nature as a vital, dynamic tool that could help improve social welfare in an age of globalization, including animal welfare.[17]

Consider the regulatory steps taken by the US to protect dolphins during the 1990s. In response to public outrage about the mass death of dolphins caused by common methods of fishing for tuna, the US banned imports of tuna sourced by certain fishing methods. The US' efforts were soon after crushed at the WTO, where it was accused of protectionism.[18] Though we can argue at length about the underlying motive of the US, what is important about this dispute is that it led to the creation of the International Dolphin Conservation Program (IDCP). In 1999, the US brought together Belize, Colombia, Costa Rica, Ecuador, El Salvador, the EU, Guatemala, Honduras, Mexico, Nicaragua, Panama, Peru, Vanuatu, and Venezuela to join the Agreement on the International Dolphin Conservation Program (AIDCP), whose declared objective is to eliminate dolphin mortality.[19] Similarly, the EU's efforts to ban importation of furs made from animals caught in leghold traps resulted in the US and Canada entering a common agreement with the EU and raising their standards on trapping.[20] Though one may oppose extraterritorial jurisdiction on various grounds, it can manifestly prompt states to adopt better laws for animals. If we wanted to pursue a similar strategy to prohibit trophy hunting, what would extraterritorial jurisdiction in this context look like? And how can it be exercised without causing conflict within the international community?

---

[16]Bowett, 'Jurisdiction: Changing Patterns of Authority over Activities and Resources' 1982, 14. *See also* Jennings/ Watts (eds), *Oppenheim's International Law Vol. I* 1992, 457.

[17]Schiff Berman, *Global Legal Pluralism* 2012, 237; Cover, 'The Uses of Jurisdictional Redundancy' 1981, 639-682.

[18]*See* WTO, *US – Restrictions on Imports of Tuna,* Report of the Panel of 3 September 1991, WT/DS21/R - 39S/155 (not adopted); *US – Restrictions on Imports of Tuna,* Report of the Panel of 16 June 1994, WT/DS29/R (not adopted).

[19]International Dolphin Conservation Program Agreement (AIDCP), 5 May 1998, 1999 OJ (L 132) 3.

[20]While the agreement between the EU and Canada is binding, the agreement between the EU and the US solely incorporates a pledge to promote 'humane' standards of trapping: Agreement on International Humane Trapping Standards between the European Community, Canada and the Russian Federation, 15 December 1997, 1998 O.J. (L 42) 43; U.S.-EU Agreed Minute on Humane Trapping Standards, 1998 O.J. (L 219) 26, at 4.

5   Extraterritorial Jurisdiction: Mapping the Options

Extraterritorial jurisdiction is a generally recognized and accepted regulatory tool in criminal, human rights, environmental, labour, antitrust, securities, and banking law.[21] In animal law, by contrast, extraterritorial jurisdiction is still largely unexplored.[22] Let us therefore, in the following, sketch possible forms of extraterritorial jurisdiction at the example of trophy hunting.

The doctrine of jurisdiction distinguishes territorial, indirect extraterritorial and direct extraterritorial jurisdiction.[23] Territorial jurisdiction regulates domestic affairs, for example, by prohibiting trophy hunting on domestic territory. Indirect extraterritorial laws also regulate domestic affairs but have an ancillary effect on foreign territory. Among those norms are import restrictions of trophies intended to protect a society from participating in despised practices through consumption; these norms may (or may not) *en passant* protect animals abroad. Finally, a state exercises direct extraterritorial jurisdiction when it regulates a state of fact abroad, namely by directly prohibiting the hunting of animals on foreign territory. It can do so by invoking such principles of international law as the active personality, the passive personality, the subjective territoriality, or the effects principle of jurisdiction. Here, I outline these means of direct extraterritorial jurisdiction for animal law, beginning with the *lex lata*.

5.1   Lex Lata *Options for Regulating Trophy Hunting*

Active Personality Principle   Under international law, the active personality principle gives states the right to prescribe actions of their nationals abroad. The principle is the most accepted and universally used basis for extraterritorial jurisdiction, as it relies on a loyalty connection between a state and its nationals.[24] In recent years, state practice has extended the principle to residents and domiciliaries operating abroad, where there is a strong enough connection between them and their home

---

[21] See for a general overview, Ryngaert, *Jurisdiction in International Law* 2015; Scott, 'Extraterritoriality and Territorial Extension in EU Law' 2014, 87-126; Zerk, 'Extraterritorial Jurisdiction' 2010.

[22] See however, Blattner, *The Extraterritorial Protection of Animals* 2019.

[23] Meng, *Extraterritoriale Jurisdiktion* 1994, 10-13; Rudolf, 'Territoriale Grenzen der staatlichen Rechtsetzung' 1973, 9-10.

[24] Meng, *Extraterritoriale Jurisdiktion* 1994, 53. Nationality represents a loyalty connection between the state and its citizen that engenders mutual rights and duties. A state protects its citizens (diplomatically and socially, grants them the right to entry, or the right to vote) and, in turn, demands their subjection to its jurisdiction. *See e.g.,* International Covenant on Civil and Political Rights, 16 December 1966, 999 UNTS 171, art. 12(4) para. 4.

states.[25] A state can use the active personality principle to prohibit its nationals or residents from hunting certain or all animals, if these acts of hunting are also prohibited on domestic territory. Importantly, because double criminality for trophy hunting is not required under international law,[26] the US could in the Cecil case prohibit Palmer from hunting endangered animals abroad regardless of whether these countries also prohibit, or even regulate these acts. This principle is thus a highly effective means to close regulatory gaps that plague animal law.

Objective Territoriality Principle The international community has also responded to the inadequacy of purely territorial jurisdiction by establishing the subjective and objective territoriality principles. If acts or omissions occur only partly in the territory of a state, the principles of subjective and objective territoriality cover the entire act or omission. The subjective territoriality principle establishes jurisdiction over an act that commenced in the territory of the state exercising jurisdiction. The complementary principle of objective territoriality gives the state in which the act was completed the right to exercise its jurisdiction over the entire act.[27] A state that wants to help end illegal trophy hunting abroad can invoke the objective territoriality principle. Since a constituent component of trophy hunting is trophy display at home, it is reasonable to argue that the act of trophy hunting is completed by the act of importation: importing the trophy is a constituent element of the crime, which is consummated in the US.

This line of argument might remind some readers of transporting rules, and specifically of the case *Zuchtvieh-Export GmbH v. Stadt Kempten*. In this case, the European Court of Justice (ECJ) held that harmonized provisions on the transport of animals destined for exports outside the EU apply beyond EU territory.[28] Zuchtvieh-Export GmbH addressed the Court in matters concerning a decision by the Stadt Kempten, whereby it refused clearance for a consignment of cows to be transported to Andijan (Uzbekistan). The Court sided with Kempten, holding that from the point of departure to the point of destination in any third country, the organizer of the journey must abide by Council Regulation (EC) No. 1/2005, by providing the necessary information on watering and feeding intervals, journey times, and resting periods.[29] These duties, as the Court clarified, are due during all stages of the journey, whether they take place inside the territory of the EU or in the territory of

---

[25]For example, under French law, sexual intercourse with minors abroad is punishable based on the habitual residence of the perpetrator: France, Code pénal, 19 December 2015, art. 227-27-1.

[26]Bantekas, 'Criminal Jurisdiction of States under International Law' 2011, 13.

[27]Crawford, *Brownlie's Principles of Public International Law* 2012, 458; Inazumi, *Universal Jurisdiction in Modern International Law* 2005, 22; Harvard Research in International Law, 'Jurisdiction with Respect to Crime' 1935, 484-94.

[28]ECJ, *Zuchtvieh-Export GmbH v. Stadt Kempten*, Judgment of 23 April 2015, Case C-424/13, 2015 E.C.R. I-1251.

[29]Council Regulation 1/2005 on the Protection of Animals During Transport and Related Operations and Amending Directives 64/432/EEC and 93/119/EC and Regulation 1255/97/EC, 2005 O.J. (L 3) 1, and corrigendum 2011 O.J. (L 336) 86.

third countries. To justify the Regulation's extraterritorial application, the Court argued that animal welfare is a legitimate objective and public interest enshrined in art. 13 TFEU and in art. 14(1)(a)(ii) and (b) of Regulation No. 1/2005 that must be respected even outside EU borders. It therefore seems that the Court qualified the transport as an export over which the EU had control *qua* its public morals.

While the Court's justification is certainly understandable in the context of trade, it failed to note the crucial difference between transporting rules and export control laws. Export controls allow or disallow exports based on the laws of the destination country and extend beyond the transportation process. In contrast, laws on transport do not purport to regulate animal welfare beyond the point of arrival; they are an application of the subjective and objective territoriality principles. This difference is relevant because laws governing export controls are much more delicate, legally, than norms based on accepted jurisdictional principles.[30] Rather than risk venturing into a heated political debate, the ECJ could have chosen an easy and more coherent strategy by invoking the subjective territoriality principle, which would have given it full jurisdiction over cross-border animal transports.

Though states have not yet entertained this line of argument, it promises to successfully address and solve a considerable portion of cross-border issues in animal law. As states increasingly rely on the subjective and objective territoriality principles to combat business crime, corruption, and cross-border financial crimes, an extension of the principle to animal law seems only coherent.

Ordre Public Exception in Private International Law By paying large sums to hunt animals abroad, foreign nationals are concluding a private contract with park rangers domiciled in the target country. If either of the parties does not fulfil their contractual obligations, the other party can sue. According to the general contracts rule, the courts of the state where an obligation should have been performed have jurisdiction. The contract over trophies may be twofold, encompassing both the act of killing the animal and importing the trophy to the hunter's home country. According to the Brussels regime, which is representative of the rules in most private international law systems, a sales contract's place of performance is the place where the goods should have been delivered,[31] that is, the US in Cecil's case. Even if a US court took jurisdiction, however, it is likely that the court would apply foreign law, because the Rome I Regulation gives parties the choice of law or applies the law of

---

[30]*See* the dismissive stance of the Court in ECJ, *The Queen and Minister of Agriculture, Fisheries and Food, ex parte Compassion in World Farming Limited*, Judgment of 19 March 1998, Case C-1/96, 1998 ECR I-1251, paras. 66-69. In 1998, the ECJ was called by Compassion in World Farming (CIWF) and the International Fund for Animal Welfare (IFAW) to declare that the UK was entitled to ban exports of calves that would prospectively be confined outside its territory in veal crates, a method widely criticized for disregarding the most fundamental interests of calves. The ECJ held that member states were barred from invoking article 36 of the Treaty Establishing the European Community to rely on public morality, public policy, or the protection of the health or life of animals to justify export restrictions.

[31]Council Regulation 1215/2012 on Jurisdiction and the Recognition and Enforcement of Judgments in Civil and Commercial Matters (Recast), 2012 OJ (L 351) 1, art. 7 para. 1 lit. b.

the seller's domicile,[32] in our case Zimbabwe. Under US law, a court has a larger margin of appreciation to enter the claim and apply its own law, based on its distinct 'most significant relationship' doctrine, which precludes party choice.[33] If foreign law is applied nonetheless, it likely leads to the result that trophy hunting is considered legal. The only way to avert such a judgment is to invoke the *ordre public* exception, i.e., showing that the application of foreign law would be manifestly incompatible with a home state's public policy.[34] A strong indication for the assumption that the US should be able to invoke this exception in Cecil's case is that 74% of the population opposes canned hunting, i.e., hunting an animal raised on a game ranch in a confined area.[35] An important caveat for applying the public order exception is, however, that the act contravening fundamental national values must also be prohibited on domestic territory. In this case, the US fulfils this criterion by section 9(a)(B) ESA, which prohibits hunting endangered animals within the US.

5.2   Lex Ferenda *Options for Regulating Trophy Hunting*

This brief overview illustrates the various *lex lata* options available to states that want to combat trophy hunting. Yet these options do not cover all jurisdictional interests of states, and there are good reasons to argue that the existing catalogue of jurisdictional options should be expanded to protect animals more effectively across the border.[36] In the following, I take a critical positivist approach to exploring novel applications of the effects principle and the universality principle, with this end in mind.

Effects Principle   Based on the effects principle, a state can exercise jurisdiction over activities outside its territory if these activities have or threaten to have a substantial effect on domestic territory.[37] The effects principle historically emerges

---

[32]Regulation 593/2008 of the European Parliament and of the Council on the Law Applicable to Contractual Obligations (Rome I), 2008 O.J. (L 177) 6, art. 4(1).

[33]American Law Institute, 'Restatement (Second) of Conflict of Laws of the United States' (Philadelphia: The American Law Institute 1971), available at: https://www.ali.org/projects/show/conflict-laws/, § 6 cmt. c.

[34]International Law Association, 'International Civil Litigation for Human Rights Violations, Final Report' (London: ILA 2012), available at file:///Users/charlotteblattner/Downloads/Conference%20Report%20Sofia%202012.pdf, 25.

[35]Humane Society of the United States (HSUS), 'New Poll Reveals Majority of Americans Oppose Trophy Hunting Following Death of Cecil the Lion', 2015, available at: http://www.humanesociety.org/news/press_releases/2015/10/poll-americans-oppose-trophy-hunting-100715.html.

[36]For example, if antitrust law gave rise to the effects principle, it is reasonable to argue that animal law could similarly give rise to a new jurisdictional principle, or considerably alter existing ones.

[37]American Law Institute, 'Restatement (Third) of the Foreign Relations Law of the United States' (Philadelphia: The American Law Institute 1987), available at: https://www.ali.org/projects/show/foreign-relations-law-united-states/, § 402, cmt. d. It is argued the principle was accepted in the *Lotus* case: Fox, 'Jurisdiction and Immunity', in Lowe/Fitzmaurice (eds) 1996, 212.

from the objective territoriality principle[38] but is now recognized as a distinct jurisdictional principle that is chiefly used in antitrust law and which covers economic effects such as financial losses.[39] In the past years, the principle has been expanded to cover other types of effects, including environmental effects (environmental pollution, loss of biodiversity, etc.)[40] or reputational effects (relied on generally in cases of human rights violations,[41] and in the context of corruption and sex tourism).[42]

The latter variant of the effects principle could profitably be used to regulate trophy hunting across borders. A home country could in this sense prohibit hunting animals abroad, if its reputation is damaged by these practices. Transforming the effects principle in this manner, however, yields potential for abuse. Reputations, values, and sensitivities vary widely across states. What one state perceives as offensive, another does not. States could also easily end up imposing their public morals in a disproportionate and illegitimate way to other cultures or nations, which more likely threatens international peace. Zerk accordingly argues that this kind of effects-based extraterritorial jurisdiction would not stand a chance in international law.[43] The only way the international community might be persuaded to accept this version of the effects principle is by restricting the scope of its application.

As with the ordinary effects principle, the reputational effect sustained by the home country could be limited to substantial effects (i.e., shared by a majority of its citizens) that are directly felt at home, and were reasonably foreseeable to the violator.[44] The state invoking the principle would also need to show it is more

---

[38]Bantekas, 'Criminal Jurisdiction' 2011, 5; Crawford, *Brownlie's Principles* 2012, 459.

[39]*See e.g.,* Ryngaert, *Jurisdiction over Antitrust Violations in International Law* 2008.

[40]A 2012 study of the European Union Directorate-General for External Policies found the principle applies to environmental law based on environmental effects (environmental pollution, loss of biodiversity, etc.): Directorate-General for External Policies, Policy Department, 'The Extraterritorial Effects of Legislation and Policies in the EU and US, requested by the European Parliament's Committee on Foreign Affairs', 2012, available at: http://www.europarl.europa.eu/the-secretary-general/en/organisation/directorate-general-for-external-policies-of-the-union, 5.

[41]In labour law, the effects are reputational. States resent being identified with domestic parent corporations of enterprises that run on cheap labour, forced labour, or human rights violations abroad (The Parliament of the Commonwealth of Australia (Parliamentary Joint Statutory Committee on Corporations and Securities), Report on the Corporate Code of Conduct Bill 2000 (Parliament House, Canberra, June 2001)). Similarly, in *Kiobel*, a minority opinion argued that foreign human rights violations should be remedied domestically, because they 'substantially and adversely affect [...] an important American national interest.' (US, Kiobel v. Royal Dutch Petroleum, 133 S.Ct. 1659, 1671 (2013) (Breyer, J., concurring)).

[42]Corruption and sex tourism threaten a state's international reputation, which is why domestic law is frequently applied to these extraterritorial events: Zerk, 'Extraterritorial Jurisdiction', 2010, 207-8.

[43]Zerk, *Multinationals and Corporate Social Responsibility* 2008, 110-111.

[44]*See e.g.,* US, *Hartford Fire Insurance Co v. California*, 509 U.S. 764 (1993); US Department of Justice, 'Antitrust Enforcement Guidelines', 1995, available at: https://www.justice.gov/atr/anti trust-enforcement-guidelines-international-operations, paras. 3.1, 3.12; Commission Notice, Guidelines on the Effect on Trade Concept Contained in arts. 81 and 82 of the Treaty (2004/C 101/07),

affected than any other state. Reputational damage might occur, e.g., when animals abused abroad were transported there from the affected country, or when a former domestic corporation of the affected state now conducts abhorrent animal experiments abroad, or in any case where there is substantial proximity to the state exercising jurisdiction. The New Zealand Animal Welfare Act in this sense reads '[t]he purpose of this Part is to *protect the welfare of animals being exported* from New Zealand and to *protect New Zealand's reputation* as a responsible exporter of animals [. . .].'[45]

Universality Principle Under international law, the universality principle endows states with prescriptive jurisdiction over egregious crimes, regardless of where or by whom they were committed. Its legitimacy emanates from the fact that certain crimes are so serious and threatening that all states share an interest in preventing or stopping them.[46]

The universality principle could be fruitfully employed to combat the most egregious crimes against animals—crimes strongly condemned by the international community. An absolute majority of states expressly recognizes that animals are sentient beings to whom we owe moral and legal duties. Anti-cruelty laws of many states are based on the idea that it is abhorrent to cause physical and psychological harm to animals or to deprive them of basic needs. An overwhelming majority of states has also enshrined the obligation to treat animals humanely and to spare them unnecessary suffering. These laws serve as proof of a universal belief that animals be properly treated: the general principle of animal welfare in international law.[47] Scholars predict this principle will develop into a norm of customary international law, concomitant with rising global concerns for animals and the on-going juridification of animal law.[48] If this proves true, states could criminalize animal cruelty and suffering that undermine fundamental values of humanity and are condemned by the world community wherever and by whomever they are committed.

The universality principle also covers crime that is not necessarily the most heinous, but which is detached from states' jurisdictions, such as piracy.[49] States could prosecute crimes against animals, if those crimes manifestly escape the

---

2004 O.J. (C 101) 81, para. 92 (substantial), para. 24 (direct or indirect), para. 23 (foreseeable). Ryngaert qualifies the test of direct, substantial, and reasonably foreseeable effects as a norm of customary international law: Ryngaert, *Jurisdiction over Antitrust Violations* 2008, 58.

[45]New Zealand, Animal Welfare Act 1999, Public Act 1999 No 142, §38 (emphasis added).

[46]ICJ, *Arrest Warrant (Dem. Rep. Congo v. Belg.)*, Judgment of 14 February 2002, ICJ Reports 2002, 81 (Joint Separate Opinion of Judges Higgins, Kooijmans, and Buergenthal).

[47]Blattner, 'An Assessment of Recent Trade Law Developments from an Animal Law Perspective' 2016, 302; Bowman/Davies/Redgwell, *Lyster's International Wildlife Law* 2010, 678 f.; Sykes, 'Sealing Animal Welfare into the GATT Exceptions' 2014, 471-498; Trent/Edwards/Felt/O'Meara, 'International Animal Law, with a Concentration on Latin America, Asia, and Africa' 2005, 77.

[48]Bowman/Davies/Redgwell, *Lyster's* 2010, 680; Sykes, 'Sealing Animal Welfare' 2014, 479-80.

[49]United Nations Convention on the Law of the Sea, 10 December 1982, 1833 UNTS 3, art. 101(a); US Third Restatement 1987 (n. 37), §404.

jurisdictional authority of most states. Especially if animal exploitation coincides with organized crime—as is often the case with trophy hunting and illegal wildlife trade[50]—states should be entitled to expand their universal jurisdiction to ensure that those crimes will not go unpunished.

6   Trophy Hunting Is Only the Tip of the Iceberg

Trophy hunting is only one of many more cases in which the current inadequacy of international animal law manifests itself. Though most consumers like to think that animals used for agricultural production enjoy a high standard of care and are mainly produced 'at home', these animals are readily transported, shipped, and flown across states to save on production costs. To meet growing demand for animal products and save land and labor costs, corporations have merged into multinationals and split up production across sites in the territories of different countries. Shrimp, for instance, are harvested in the North Sea and driven 2000 miles south to Morocco, where producers profit from cheap labour. After they are shelled and enriched with preservatives to inhibit decay, they are transported back to Northern Europe.[51] For other products, including meat, eggs, milk, and compound products derived from them, another couple of production steps in different states might add to this.

On-going division of labor, fewer barriers of trade, and foreign direct investments also encourage companies to disperse production over the globe. In the coming years, we anticipate a wave of agricultural outsourcing from the minority world to the majority world, prompted by heavy investments in farmland in the majority world.[52] This is expected to be the third wave of global industry outsourcing, following the first wave of manufacturing outsourcing in the 1980s and the second wave of information outsourcing in the 1990s. Relocation and outsourcing are also common in the research industry, notably among biomedical and pharmaceutical institutions and their supplying facilities.[53] Overall, multinational corporations, which own most of the world's domesticated animals,[54] are highly mobile and do not shy away from moving production to states with more lenient regulatory environments.

---

[50]United Nations Office on Drugs and Crime, Bulletin on Organised Crime in Southern Africa (2012), available at: https://www.unodc.org/documents/southernafrica/Bulletin_on_organised_crime_in_Southern_Africa/UNODC_ROSAF_-_Bulletin_on_Organised_Crime_in_Southern_Africa_-_Issue_1.pdf, i.

[51]Documentary Presseportal, 'Vorsicht Krabbe! – Das grosse Geschäft mit dem kleinen Tier', 2014, available at: http://presse.phoenix.de/dokumentationen/2014/10/20141017_Krabbe/20141017_Krabbe.phtml.

[52]'Outsourcing's Third Wave', *The Economist* (21 May 2009).

[53]Laster, 'Plan to Breed Lab Monkey Splits Puerto Rican Town' 2009; Pocha, 'Outsourcing Animal Testing: US Firm Setting Up Drug-Trial Facilities in China' 2006.

[54]Park/Singer, 'The Globalization of Animal Welfare', 2012, 122-133.

Given these developments, it can confidently be said that issues of animal production and protection have become so globally entangled that jurisdictional connections often cannot be traced to a single state anymore. This approximation has brought states' regulatory particularities more sharply into focus, by accentuating remaining differences in regulation.[55] This, in turn, makes it convenient for corporations to choose home states based on the regulatory advantages they provide them, which stokes fear among states that business will move somewhere more advantageous. Rather than autonomously exercising their sovereign authority, states have begun to compete with each other through their regulatory systems, and learned that they gain a comparative advantage by designing their laws to the investors' and producers' liking.[56] These dynamics are commonly described as regulatory competition, also known as jurisdictional competition or systems competition.[57]

As states compete over jurisdictional authority, regulation tends to converge towards laxity. Corporations predictably seek to maximize capital value, which is more likely when governments intervene less, and when corporations incur fewer costs than their competitors in other jurisdictions do.[58] To attract corporations and gain 'regulatory market shares', states lax their legal standards and create incentive for other states to follow suit. This competitive move eventually results in global convergence toward a lower common denominator, also known as competition in laxity or the race to the bottom.[59]

Animal law theorists often argue that regulatory competition in animal law moves towards laxity.[60] The more rigidly laws insist on specific performances of corporations—such as by determining how animals ought to be bred, reared, transported, or slaughtered—the more corporations are disabled from choosing the cheapest factors of production needed to outpace competitors. Because policies that seek to improve animal welfare commonly restrict business activities, these standards tend to impede market growth.[61] And vice versa, because there is always an economically more

---

[55]Kaufmann, *Globalisation and Labour Rights* 2007, 232 *et seq.*; Picciotto, 'The Regulatory Criss-Cros' 1996, 89-123.

[56]Bratton/McCahery/Picciotto/Scott, 'Regulatory Competition and Institutional Evolution' 1996, 2; Koenig-Archibugi, 'Global Regulation' 2010, 413; Murphy, *The Structure of Regulatory Competition* 2004, 4.

[57]Eidenmüller argues that systems competition must be differentiated from regulatory competition. Systems competition is a competition not only of legal rules but also of a state's infrastructure, while regulatory competition refers to the competition of laws only: Eidenmüller, 'The Transnational Law Market, Regulatory Competition, and Transnational Corporations,' 2011, 715.

[58]Kaufmann, *Globalisation and Labour Rights* 2007, 15; Murphy, *The Structure of Regulatory Competition* 2004, 10-11.

[59]The race to the bottom is also frequently called 'Delaware effect' (coined by Vogel, *Trading up: Consumer and Environmental Regulation in a Global Economy* 1995) or 'Zug effect' (Murphy, *The Structure of Regulatory Competition* 2004, 6).

[60]Kelch, 'Towards Universal Principles for Global Animal Advocacy' 2016, 82.

[61]Vernon and Nwaogu argue that a number of recently introduced changes to the regulatory framework of the EU have the potential to act as a barrier for future innovation. These barriers include, in particular, testing and marketing bans of cosmetic products: Vernon/Nwaogu,

efficient jurisdiction where capital can move, unhampered regulatory competition frustrates the successful introduction of or adherence to well-established levels of animal law.

In other cases, industries merely threaten to relocate their production to prevent parliamentary or ballot initiatives from improving the legal status of farm, research, or wild animals. Such threats are often enough to prompt states to enter a state of regulatory chill, meaning they decline to raise animal protection standards.[62] For example, in 2015, the German public pushed for a national ban on chick shredding, but the *Bundestag* feared that the ban might prompt hen producers to relocate their facilities to less regulated countries.[63] Eventually, parliament did not adopt the ban and overruled the people's will to save 50 million male chicks per year from being shredded alive in the first few minutes of their lives. Regulatory chill—like competition in laxity—thus often defies societal demands and new scientific evidence about the complex and valuable lives of animals.

States that seek to withstand this pull towards laxity and decide to adopt stricter animal laws are often penalized. In 2006, the US tried to ban the commercial slaughter of horses for meat by prohibiting the issuance of federal funds to inspectors of horsemeat.[64] Without federal meat inspections, institutions that slaughtered horses could not run their businesses legally. Within a year of the ban, horse exports from the US to Mexico increased by 312%.[65] In other words, the entire horse slaughter industry of the US was effectively outsourced to Mexico, and this reignited societal concerns about animal welfare. The only way states can successfully counter this disconcerting development is to use the principles of extraterritorial jurisdiction outlined herein.

7   Concluding Remarks

By and large, animals lack a voice in the formation of law and have no opportunity to escape oppressive jurisdictional authority. Most states use their territorial primacy to attract foreign investment by bereaving animals—who are at the mercy of a single

---

'Comparative Study on Cosmetics Legislation in the EU and Other Principal Markets', Final Report Contract No. FIF.20030624, Prepared for European Commission DG Enterprise (Norfolk 2004), available at: http://rpaltd.co.uk/uploads/report_files/j457-final-report-cosmetics.pdf, 37.

[62] Analogously: Murphy, *The Structure of Regulatory Competition* 2004, 7.

[63] Deutscher Bundestag, 18. Wahlperiode, Gesetzesentwurf des Bundesrates zur Änderung des Tierschutzgesetzes, Drucksache 18/6663, 11 November 2015, Stellungnahme der Regierung, 10-11.

[64] US, The Agriculture, Rural Development, Food and Drug Administration, and Related Agencies Appropriations Act, 2006, 119 U.S. 2120, Public Law 109-97, H.R. 2744-45, §794. The ban was upheld in US, The Agriculture, Rural Development, Food and Drug Administration, and Related Agencies Appropriations Act, 2014, 128 U.S. 5, Public Law 113-76, H.R. 3547.

[65] Nolen, 'U.S. Horse Slaughter Exports to Mexico Increase 312%' 2008.

regulator—of protection and rights. This praxis is increasingly criticized by citizens witnessing the 'globalization of animal cruelty'[66] and has brought jurisdictional issues to the forefront of the discussion in animal law. Ideally, these problems would be addressed and solved by concluding an international treaty in animal law. Yet, an economic analysis shows that this strategy is unlikely to work, and perhaps not even desirable, because international agreements tend to cap law at the lowest common denominator.

Extraterritorial animal law may offer a way out of this dilemma. The example of trophy hunting shows the range of possibilities the international doctrine of jurisdiction provides: the active personality principle, the subjective and objective territoriality principle, the *ordre public* exception, the effects principle, and the universality principle. By adopting these, we could abandon the archaic territorial conception of jurisdiction that binds individuals to it in an exclusive fashion and fences off other sovereigns. The territorial primacy a state may once have enjoyed vis-à-vis its regulatees offered ample room for misuse by bereaving regulatees—who are at the mercy of this single regulator—of protection and welfare. Thanks to the development of the modern law of jurisdiction, states can choose among viable jurisdictional options to protect animals abroad. These options are especially valuable to animals—more than to any other group that benefits from extraterritorial jurisdiction—because they still live under a totalitarian regime of law. Needless to say, extraterritorial jurisdiction runs the risk of being used to oppress or discriminate others, but if properly applied and strengthened with the necessary safety valves, it can be a powerful tool to advance our ongoing struggle for interspecies justice.

References

Bantekas, I. (2011). Criminal jurisdiction of states under international law. In R. Wolfrum (Ed.), *Max Planck encyclopedia for public international law*. Oxford: Oxford University Press online ed.

Blattner, C. (2016). An assessment of recent trade law developments from an animal law perspective: Trade law as the sheep in the Wolf's clothing? *Animal Law, 22*(2), 277–305.

Blattner, C. (2019). *The extraterritorial protection of animals*. Oxford: Oxford University Press.

Bowett, D. W. (1982). Jurisdiction: Changing patterns of authority over activities and resources. *British Yearbook of International Law, 53*, 1–26.

Bowman, M., Davies, P., & Redgwell, C. (2010). *Lyster's international wildlife law* (2nd ed.). Cambridge: Cambridge University Press.

Bratton, W. W., McCahery, J., Picciotto, S., & Scott, C. (1996). Regulatory competition and institutional evolution, in international regulatory competition and coordination. In W. W. Bratton, J. McCahery, S. Picciotto, & C. Scott (Eds.), *International regulatory competition and coordination* (pp. 1–55). Oxford: Clarendon Press.

Capecchi, C., & Rogers, K. (2015, July 29). Killer of Cecil the Lion finds out that he is a target now of internet vigilantism. *New York Times*.

---

[66]White/Cao, 'Introduction: Animal Protection in an Interconnected World' 2016, 2.

Cover, R. (1981). The uses of jurisdictional redundancy: Interest, ideology, and innovation. *William & Mary Law Review, 22*, 639–682.

Crawford, J. (2012). *Brownlie's principles of public international law* (8th ed.). Oxford: Oxford University Press.

Eidenmüller, H. (2011). The transnational law market, regulatory competition, and transnational corporations. *Indiana Journal Global Legal Studies, 18*, 707–749.

Estes, J., Crooks, K., & Holt, R. D. (2013). Ecological role of predators. In S. A. Levin (Ed.), *Encyclopedia of Biodiversity* (Vol. 4, 2nd ed., pp. 229–249). Amsterdam: Elsevier.

Ferraro, P. J., McIntosh, C., & Ospina, M. (2007). The effectiveness of the US Endangered Species Act: An econometric analysis using matching methods. *Journal of Environmental Economics and Management, 54*, 245–261.

Fox, H. (1996). Jurisdiction and immunity. In V. Lowe & M. Fitzmaurice (Eds.), *Fifty years of the international court of justice: Essays in Honour of Sir Robert Jennings* (pp. 210–236). Cambridge: Cambridge University Press.

Guzman, A. T. (1998). Is international antitrust possible? *New York University Law Review, 73*, 1501–1548.

Harvard Research in International Law. (1935). Jurisdiction with respect to crime. *American Journal of International Law, 29*, 435–651.

Inazumi, M. (2005). *Universal jurisdiction in modern international law: Expansion of national jurisdiction for prosecuting serious crimes under international law.* Antwerp: Intersentia.

Jennings, R., & Watts, A. (Eds.). (1992). *Oppenheim's international law* (Vol. I, 9th ed.). Essex: Longman Group UK Limited.

Kaufmann, C. (2007). *Globalisation and labour rights: The conflict between core labour rights and international economic law.* Portland: Hart.

Kelch, T. (2016). Towards universal principles for global animal advocacy. *Transnational Journal of Environmental Law, 5*, 81–111.

Koenig-Archibugi, M. (2010). Global regulation. In R. Baldwin, M. Cave, & M. Lodge (Eds.), *The Oxford handbook of regulation* (pp. 407–437). Oxford: Oxford University Press.

Laster, J. (2009, November 30). Plan to Breed Lab Monkey Splits Puerto Rican Town. *US News & World Report.*

McOmber, E. M. (2002). Problems in enforcement of the convention on international trade in endangered species. *Brooklyn Journal of International Law, 27*, 674–701.

Meng, W. (1994). *Extraterritoriale Jurisdiktion im öffentlichen Wirtschaftsrecht.* Berlin: Springer.

Murphy, D. D. (2004). *The structure of regulatory competition: Corporations and public policies in a global economy.* Oxford: Oxford University Press.

Nolen, R. S. (2008, January 15). U.S. Horse Slaughter Exports to Mexico Increase 312%. *AVMA News.*

Onishi, N. (2015, August 10). Outcry for Cecil the Lion could undercut conservation efforts. *New York Times.*

Packer, C., Brink, H., Kissui, B. M., Maliti, H., Kushnir, H., & Caro, T. (2011). Effects of trophy hunting on lion and leopard populations in Tanzania. *Conservation Biology, 25*, 142–153.

Park, M., & Singer, P. (2012). The globalization of animal welfare: More food does not require more suffering. *Foreign Affairs, 91*, 122–133.

Picciotto, S. (1996). The regulatory criss-cross: Interaction between jurisdictions and the construction of global regulatory networks. In W. W. Bratton, J. McCahery, S. Picciotto, & C. Scott (Eds.), *International regulatory competition and coordination* (pp. 89–123). Oxford: Clarendon Press.

Pocha, J. S. (2006, November 25). Outsourcing animal testing: US firm setting up drug-trial facilities in China: Where scientists are plentiful but activists aren't. *Boston Globe.*

Punch, S. (2016). Exploring children's agency across majority and minority world contexts. In F. Esser, M. S. Baader, T. Betz, & B. Hungerland (Eds.), *Reconceptualizing agency and childhood: New perspectives in childhood studies* (pp. 183–196). London: Routledge.

Rudolf, W. (1973). Territoriale Grenzen der staatlichen Rechtsetzung. *Berichte der Deutschen Gesellschaft für Völkerrecht, 11*, 7–46, at 9-10.

Ryngaert, C. (2008). *Jurisdiction over antitrust violations in international law*. Oxford: Oxford University Press.

Ryngaert, C. (2015). *Jurisdiction in international law* (2nd ed.). Oxford: Oxford University Press.

Schiff Berman, P. (2012). *Global legal pluralism: A jurisprudence of law beyond borders*. Cambridge: Cambridge University Press.

Scott, J. (2014). Extraterritoriality and territorial extension in EU law. *American Journal of Comparative Law, 62*, 87–126.

Sykes, K. (2014). Sealing animal welfare into the GATT exceptions: The international dimension of animal welfare in WTO disputes. *World Trade Review, 13*, 471–498.

Trent, N., Edwards, S., Felt, J., & O'Meara, K. (2005). International animal law, with a concentration on Latin America, Asia, and Africa. In D. J. Salem & A. N. Rowan (Eds.), *The state of the animals III* (pp. 65–77). Washington: Humane Society Press.

Vernon, J., & Nwaogu, T. A. (2004). *Comparative study on cosmetics legislation in the EU and other principal markets* (Final Report Contract No. FIF.20030624). Prepared for European Commission DG Enterprise (Norfolk 2004), Available at http://rpaltd.co.uk/uploads/report_files/j457-final-report-cosmetics.pdf

Vogel, D. (1995). *Trading up: Consumer and environmental regulation in a global economy*. Cambridge: Harvard University Press.

White, S., & Cao, D. (2016). Introduction: Animal protection in an interconnected world. In S. White & D. Cao (Eds.), *Animal law and welfare: International perspectives* (pp. 1–11). Cham: Springer.

Zerk, J. A. (2010). Extraterritorial jurisdiction: Lessons for the business and human rights sphere from six regulatory areas (Corporate Social Responsibility Initiative Working Paper No. 59, Harvard University 2010), Available at https://sites.hks.harvard.edu/m-rcbg/CSRI/publications/workingpaper_59_zerk.pdf

Zerk, J. A. (2008). *Multinationals and corporate social responsibility: Limitations and opportunities in international law*. Cambridge: Cambridge University Press.

Charlotte Blattner (Dr. iur., LL.M. (Harvard)) is a visiting researcher at Harvard Law School, where she works at the intersection of animal and environmental law. She is the author of the 2019 monograph *Protecting Animals Within and Across Border* (Oxford University Press) and co-editor (with Will Kymlicka and Kendra Coulter) of the collection *Animal Labour: A New Frontier of Interspecies Justice?* (Oxford University Press, 2020).

Chapter 13
Protection of Animals Through Human Rights: The Case-Law of the European Court of Human Rights

Tom Sparks

Abstract The chapter discusses the potential of a human rights framework to contribute to the growth and development of global animal law. It takes as example the jurisprudence of the European Court of Human Rights, and examine the major trends in the Court's judgments and admissibility decisions that directly or indirectly concern the rights or welfare of animals. It is concluded that the Court is not indifferent to the welfare of animals, but that animal welfare is instrumentalised: it is understood not as a good in itself, but is instead valued for its implications for human welfare and rights. The chapter then considers the obstacles that the anthropocentrism of the human rights idea and the instrumentalisation of animal concerns present to the use of human rights frameworks to further the development of global animal law, as well as the opportunities that exist in the meeting of these paradigms. It concludes that although the telos of human rights law is different from that of animal law, nevertheless there exist many overlapping concerns within which mutually beneficial interactions are possible.

1 Introduction

The institutionalisation of human rights under the regional human rights frameworks has given legal force to an idea already rich in transformative potential. Human rights have become a vital tool in efforts to achieve change both for individuals, and across legal systems. To seek, though, to harness the potential of human rights institutions and discourse to advance animal welfare and animal rights may appear farfetched: one could be forgiven for a certain scepticism concerning the availability of space for animal concerns in this deliberately human-centred area of law and policy. Nevertheless, there are indications of certain modest advances which offer

T. Sparks (✉)
Max Planck Institute for Comparative Public Law and International Law, Heidelberg, Germany
e-mail: sparks@mpil.de

© The Author(s) 2020

153

A. Peters (ed.), *Studies in Global Animal Law*, Beiträge zum ausländischen öffentlichen Recht und Völkerrecht 290,
https://doi.org/10.1007/978-3-662-60756-5_13

opportunities for animal welfare and rights concerns to be considered within the framework of human rights.

This chapter will consider the case-law of the European Court of Human Rights (ECtHR) in the field of animal law, and will identify the broad themes and trends within that jurisprudence. The ECtHR has only rarely considered questions either directly or tangentially relevant to the protection of animal rights, but there are nonetheless almost thirty relevant judgments and admissibility decisions,[1] which will be divided thematically (Sects. 2 and 3).[2] Section 4 will then consider the current legal and conceptual barriers to more effective animal protection under the Convention, and will make some tentative remarks on the potential of the ECHR and its Court (as well as human rights frameworks more broadly) to contribute to the development of global standards on animal welfare.

---

[1] The relevant cases are, in chronological order, ECHR, *Steel and Others v. UK*, Chamber Judgment of 23 September 1998, Application No. 24838/94; ECHR, *Chassagnou and Others v. France*, Grand Chamber Judgment of 29 April 1999, Applications Nos. 25088/94, 28331/95, and 28443/95; ECHR, *Bladet Tromsø and Stensaas v. Norway*, Grand Chamber Judgment of 20 May 1999, Application No. 21980/93; ECHR, *Hashman and Harrup v. UK*, Grand Chamber Judgment of 25 November 1999, Application No. 25594/94; ECHR, *Geert Drieman and Others v. Norway*, Third Section Decision on Admissibility of 4 May 2000, Application No. 33678/96; ECHR, *Cha'are Shalom Ve Tsedek v. France*, Grand Chamber Judgment of 27 June 2000, Application No. 27417/95; ECHR, *Verein gegen Tierfabriken v. Switzerland*, Second Section Judgment of 28 June 2001, Application No. 24699/94 (*VgT No. I*); ECHR, *Kyrtatos v. Greece*, First Section Judgment of 22 May 2003, Application No. 41666/98; ECHR, *Piippo v. Sweden*, Second Section Partial Decision on Admissibility of 7 December 2004, Application No. 70518/01; ECHR, *Steel and Morris v. UK*, Fourth Section Judgment of 15 February 2005, Application No. 68416/01; ECHR, *Piippo v. Sweden*, Second Section Decision on Admissibility of 21 March 2006, Application No. 70518/01; ECHR, *Schneider v. Luxembourg*, Second Section Judgment of 10 July 2007, Application No. 2113/04; ECHR, *Baudinière and Vauzelle v. France*, Third Section Decision on Admissibility of 6 December 2007, Application Nos. 25708/03 and 25719/03; ECHR, *Nilsson v Sweden*, Third Section Decision on Admissibility of 26 February 2008, Application No. 11811/05; ECHR, *Verein gegen Tierfabriken Schweiz v. Switzerland (No.2)*, Grand Chamber Judgment of 30 June 2009, Application No. 32772/02 (*VgT No. II*); ECHR, *Friend and Others v. United Kingdom*, Fourth Section Decision on Admissibility of 24 November 2009, Application Nos. 16072/06 and 27809/08; ECHR, *Jakóbski v. Poland*, Fourth Section Judgment of 7 December 2010, Application No. 18429/06; ECHR, *Berü v. Turkey*, Second Section Judgment of 11 January 2011, Application No. 47304/07; ECHR, *Georgel and Georgeta Stoicescu v. Romania*, Third Section Judgment of 26 July 2011, Application No. 9718/03; ECHR, *ASPAS and Lasgrezas v. France*, Fifth Section Judgment of 22 September 2011, Application No. 29953/08; ECHR, *Herrmann v. Germany*, Grand Chamber Judgment of 26 June 2012, Application No. 9300/07; ECHR, *Chabauty v. France*, Grand Chamber Judgment of 4 October 2012, Application No. 57412/08; ECHR, *PETA Deutschland v. Germany*, Fifth Section Judgment of 8 November 2012, Application No. 43481/09; ECHR, *Animal Defenders International (ADI) v. UK*, Grand Chamber Judgment of 22 April 2013, Application No. 48876/08; ECHR, *Tierbefreier e.V. v. Germany*, Fifth Section Judgment of 16 January 2014, Application No. 45192/09.

[2] A small number of cases fall into neither category, and are briefly mentioned in section three.

2 The Hunting Cases

In 1998, the first hunting case came before the Court. *Steel and Others* concerned a series of individuals arrested for the English common law offence of breach of the peace for acts of protest, and who had been subject to binding over orders.[3] The protest in the case of the first applicant was the disruption of a grouse shooting party and, following her refusal to accept a binding over order, she was jailed for 28 days.[4] The Court examined the complaint under the article 5 prohibition on arbitrary deprivation of liberty and as an interference with the applicant's right to free expression (article 10). In finding no violation, it noted that Ms Steel had been subjected to 'serious interferences with the exercise of her right to freedom of expression',[5] but balanced this against the 'obstruction' of the 'lawful pastime' of the hunting party and the 'risk of disorder' arising therefrom,[6] as well as the 'importance in a democratic society of maintaining the rule of law and the authority of the judiciary'.[7] It therefore held that her arrest and detention were not disproportionate interferences with her convention rights.[8]

However, in subsequent cases, the Court has asserted that a moral conviction against hunting is capable of attracting Convention protection,[9] that animal welfare is a matter of public interest,[10] and that no Convention protection of the right to hunt exists.[11] Nevertheless, it remains a mixed practice.

2.1 Hunting Under Article 1 of Protocol 1

The 1999 case of *Chassagnou v. France* concerned ten applicants, each of whom owned land in areas regulated by the *Loi Verdeille*. Under that law, all landowners whose holdings are below a certain threshold are required to pool their lands for the purposes of creating an area within which members of the relevant municipal

---

[3]ECHR, *Steel and Others v. UK* (n. 1), paras. 6-24.

[4]Ibid., para. 13.

[5]Ibid., para. 103.

[6]Ibid.

[7]Ibid., para. 107.

[8]Ibid.

[9]ECHR, *Chassagnou v. France* (n. 1), para. 114; see also ECHR, *Schneider v. Luxembourg* (n. 1), para. 80; ECHR, *Herrmann v. Germany* (n. 1), para. 80.

[10]ECHR, *Bladet Tromsø and Stensaas* (n. 1), paras. 63-64, 73; see also ECHR, *Steel and Morris v. UK* (n. 1), para. 88; ECHR, *PETA Deutschland v. Germany* (n. 1), para. 47; ECHR, *VgT No. II* (n. 1), para. 92.

[11]ECHR, *Chassagnou v. France* (n. 1), para. 113; see also ECHR, *Nilsson v. Sweden* (n. 1), 11.

hunting association (ACCA) may freely hunt. The landowners whose property forms a part of the hunting area are automatically members of the local ACCA.[12] The applicants in the case were all ethically opposed to hunting, and made unsuccessful applications to have their properties removed from the hunting areas, and themselves released from membership of the ACCAs.[13] The Court found violations of article 11 (freedom of association) and article 1 of Protocol 1 (protection of property) taken separately, and also found violations of each of these provisions when read in conjunction with the protection from discrimination in the application of the convention (article 14).

The Court accepted that the imposition of ACCA membership and the requirement to permit hunting on the applicants' land pursued a legitimate aim (it commented that 'it is undoubtedly in the general interest to avoid unregulated hunting and encourage the rational management of game stocks').[14] Nevertheless, it recognised that the applicants' ethical objections were relevant to the assessment of the proportionality of the interference. In relation to article 1 of Protocol 1 it noted that the Government's characterisation of membership of the ACCA as 'compensation' for the loss of the exclusive right to hunt (or, as the case may be, to choose not to hunt) over one's land 'is valuable only in so far as all the landowners concerned are hunters or accept hunting.'[15] It consequently found that '[c]ompelling small landowners to transfer hunting rights over their land so that others can make use of them in a way which is totally incompatible with their beliefs imposes a disproportionate burden'.[16]

Perhaps even more telling was the Court's 2012 judgment in *Chabauty v. France*, the most recent in a line of cases brought under article 1 of Protocol 1 by hunters.[17] The applicant challenged the inclusion of his land in the hunting area, but in this case because he wished privately to rent the right to hunt on his land.[18] The Court found no violation of article 1 of Protocol 1, read in conjunction with article 14, and in so doing expressly distinguished the case from *Chassagnou* and its line[19] as a result of the applicant's (lack of) ethical objections to hunting. Absent the conflict of conscience, the decision on how hunting should be regulated fell within the state's

---

[12]ECHR, *Chassagnou v. France* (n. 1), paras. 13-15, 46.

[13]Ibid., paras. 16-18, 23-24, 28-30.

[14]Ibid., para. 79.

[15]Ibid., para. 82.

[16]Ibid., para. 85. The Court has subsequently confirmed Chassagnou in ECHR, *Schneider v. Luxembourg* (n. 1); and ECHR, *Herrmann v. Germany* (n. 1).

[17]ECHR, *Piippo v. Sweden* (Second Decision, 2006) (n. 1); ECHR, *Nilsson v. Sweden* (n. 1); ECHR, *Baudinière and Vauzelle v. France* (n. 1); and ECHR, *Chabauty v. France* (n. 1).

[18]Ibid., paras. 12-17.

[19]ECHR, *Chassagnou v. France* (n. 1); ECHR, *Schneider v. Luxembourg* (n. 1); ECHR, *Herrmann v. Germany* (n. 1).

margin of appreciation, and no disproportionate interference with the right to property was found.[20]

As the line of cases culminating in *Chabauty* shows, this is no reification of the right to property. On the contrary, the Court has been inclined to give the state a wide margin of appreciation to regulate hunting.[21] Nevertheless, ethical objections to hunting are sufficient significantly to narrow the margin of appreciation, and the Court has consistently held that national regulation of hunting must make provision for the rights of opponents of hunting to use their land in ways that accord with their beliefs. Though this is only a small step towards Convention support for animal concerns, it is nevertheless noteworthy both as a protection for animal rights activists, and because it recognises opposition to hunting (and perhaps by implication concern for animals more broadly) as a politico-moral opinion capable of attracting ECHR protection.

2.2 Hunting Under Articles 10 and 11

Articles 10 and 11 have been invoked alongside the right to property as further grounds to find the obligation to accept hunting interferes with the Convention rights of conscientious objectors,[22] but have also twice been invoked in the separate context of anti-hunting protest.[23] Here the Court has seemed more reluctant to grant protection: in neither case were the restrictions of the applicants' acts of protest in defence of animal rights considered to be violations of the Convention.

Articles 10 and 11 were first invoked by an anti-hunting protester in *Steel and Others v. UK*.[24] The first applicant claimed violations of articles 10 and 11 of the Convention, but only article 10 was considered,[25] and no violation was found.[26] The Court emphasised that the applicant's protest—which involved placing herself in front of the hunters to prevent them from firing—'created a danger of serious physical injury to herself and others' and 'risked culminating in disorder and violence.'[27] It appears, therefore, to be the applicant's direct action which justified her arrest and imprisonment, a conclusion reinforced by *Geert Drieman and Others*

---

[20]Ibid., paras. 41-50, 56-57.

[21]Çoban notes that the Court's general approach to article 1 of Protocol 1 has been to 'favour[] the public interest rather than individual rights.' Çoban, *Protection of Property Rights* 2004, 257.

[22]ECHR, *Chassagnou v. France* (n. 1), para. 103, 117; ECHR, *Schneider v. Luxembourg* (n. 1), paras. 82-83.

[23]ECHR, *Steel and Others v. UK* (n. 1); ECHR, *Drieman and Others v. Norway* (n. 1).

[24]ECHR, *Steel and Others v. UK* (n. 1), paras. 6-13. See also above, sec. 2.1.

[25]The Court decided that it was not necessary to consider the application of article 11 because the complaint did not 'raise[] any issues not already examined in the context of article 10': ECHR, *Steel and Others v. UK* (n. 1), para. 113.

[26]Ibid., paras. 102-107.

[27]Ibid., para. 105.

v. Norway. Here the applicants were arrested and held on remand for actions taken to disrupt a whale hunt in Norway's exclusive economic zone.[28] The applicants claimed that their arrest and detention violated articles 10 and 11 of the Convention, and the Court accepted that these actions amounted to an interference. Nevertheless, it decided that the application was manifestly ill founded under article 35(3) of the Convention, and therefore inadmissible.[29]

The decision that the complaint in *Drieman* was manifestly ill founded and not worthy of further consideration is somewhat surprising, and seems to indicate a hostility to direct action as a form of protest. Though it was accepted by the Court that there had been an interference with articles 10 and 11, it considered it sufficiently obvious that the state's actions were proportionate that a more detailed assessment was manifestly unnecessary. Its reasoning supports two possible (non-exclusive) interpretations: that the applicants' aims did not require protection in a democratic society; or that their methods were sufficiently outrageous that states cannot be required to tolerate such conduct in defence of the right to protest. The first, it seems, played a role. The Court noted that the interference pursued the legitimate aim of 'enforc[ing ...] the rules protecting whaling',[30] and counterbalanced that remark with a finding that the protest 'forc[ed] the whalers to abandon their lawful activity'.[31] It noted, too, that the relevant conduct 'could not enjoy the same privileged protection under the Convention as political speech or debate on questions of public interests or the peaceful demonstration of opinions on such matters'.[32] Although this latter comment is more closely tied to the question of methods, taken together these statements indicate the Court's opinion that the subject of the protest did not attract a high standard of protection.[33] On the contrary,

---

[28]ECHR, *Drieman and Others v. Norway* (n. 1), 2.

[29]Ibid., 10.

[30]Ibid., 9.

[31]Ibid., 10.

[32]Ibid.

[33]The Court has implied in a series of cases that there is a hierarchy within articles 10 and 11, wherein certain subjects (those that are "political" or in the "public interest") will receive a higher level of protection than others. See ECHR, *Sunday Times v. UK*, Grand Chamber Judgment of 26 April 1979, Application No. 6538/74, 29-30; ECHR, *Lingens v. Austria*, Grand Chamber Judgment of 8 July 1986, Application No. 9815/82, paras. 34-47; ECHR, *Thorgeirson v. Iceland*, para. 60-70; Chamber Judgment of 25 June 1992, Application No. 13778/88, paras. 55-70; ECHR, *Jersild v. Denmark*, Grand Chamber Judgment of 23 September 1994, Application No. 15890/89, paras. 25-37; and contrast ECHR, *Handyside v. UK*, Grand Chamber Judgment of 7 December 1976, Application No. 5493/72, paras. 42-59; ECHR, *Wingrove v. UK*, Chamber Judgment of 25 November 1996, Application No. 17419/90, paras. 52-64, esp. 58; ECHR, *Vereinigung Bildender Künstler v. Austria*, First Section Judgment of 25 January 2007, Application No. 68354/01, paras. 26-39.

the Court privileged the economic activity of the whalers over the protest of the animal rights activists.[34]

In its assessment of the methods, too, the Court seemed ill-disposed to the direct action of the protestors.[35] The comment above contrasting the protestors' actions with 'peaceful demonstration of opinions' on questions of public interest should be read alongside the characterisation of those actions as 'a form of coercion', and 'an ultimatum'.[36] Yet unlike in *Steel and Others* there was no suggestion that the applicants' protests had created a danger to the whalers or to any other person[37]; unlike in *Kudrevičius and Others* the disruption caused to 'activities lawfully carried out by others' did not affect a large number of people, but instead only a small group[38]; and unlike in *Taranenko v. Russia* the protests did not result in violence.[39] Nevertheless, the Court in *Drieman* directly contrasted its approach in cases involving 'the peaceful demonstration of opinions on [matters of public interest]' (in which a narrow margin of appreciation is appropriate) with the facts before it, where it found that '[c]ontracting States must be allowed a wide margin of appreciation in their assessment of the necessity of taking measures to restrict such conduct.'[40]

---

[34]In this connection, it is particularly relevant that the Court observed that the protest had been taking place unimpeded for one month, and only when the protestors' activities interfered with the hunt did the authorities take action. See ECHR, *Drieman and Others v. Norway* (n. 1), 10.

[35]Fenwick et al consider that the findings in Steel and Others and the cases that followed it demonstrate that direct action protests engage article 11 in principle: Helen Fenwick/Gavin Phillipson/Alexander Williams, Texts, Cases and Materials on Public Law and Human Rights (4th ed., Abingdon: Routledge 2017), 999. However, the Court has tended to apply a very wide margin of appreciation in such cases, characterising direct action as 'reprehensible', and implying that it cannot be considered wholly 'peaceful' even when no violent action is taken: *Kudrevičius and Others v. Lithuania*, Grand Chamber Judgment of 15 October 2015, Application No. 37553/05, paras. 173-174; see also *Steel and Others* (n. 1); *G. v. Germany*, Decision by the Commission on Admissibility of 6 March 1989, Application No. 13079/87; *Lucas v. UK*, Fourth Section Decision on Admissibility of 18 March 2003, Application No. 39013/02; *Baracco v. France*, Fifth Section Judgment of 5 March 2009, Application No. 31684/05. Nevertheless the Court has held that the margin it grants in such cases 'although wide, is not unlimited' (para. 86), and it has been willing to find that a certain level of criminal sanction (in Taranenko three years' imprisonment) is disproportionate to the aim of preventing illegal protest: ECHR, *Taranenko v. Russia*, First Section Judgment of 15 May 2014, Application No. 19554/05, paras.81-97.

[36]Ibid., 10.

[37]ECHR, *Steel v. Others* (n. 1), para. 103.

[38]ECHR, *Kudrevičius and Others v. Lithuania* (n. 35), paras. 142-184, esp. 169-175.

[39]ECHR, *Taranenko v. Russia* (n. 35).

[40]Ibid.

3 Animal Welfare and Freedom of Speech

Claims under article 10 in the context of hunting have come in parallel to article 11, in cases concerning protest. There, the articles were considered to raise the same issues.[41] Freedom of speech has also been invoked separately from the freedom of association, however; both in relation to reporting on hunting,[42] and publications by animal rights groups.[43]

The first animal welfare case to raise article 10 outwith the context of protest was Bladet Tromsø *and Stensaas v. Norway*. The case was brought by the *Bladet Tromsø* newspaper and its editor, following a successful defamation suit against them for articles which reported allegations by a seal hunt inspector of cruel and illegal practices.[44] Defamation proceedings were brought by the hunters concerned sequentially against the inspector, *Bladet Tromsø* and its editor, and (unsuccessfully) against several other media outlets. The Court began its assessment with its familiar assertion of the high importance of the press, and declared that '[i]n cases such as the present one the national margin of appreciation is circumscribed by the interest of democratic society in enabling the press to exercise its vital role of 'public watch-dog' in imparting information of serious public concern'.[45] It held too that 'in order to determine whether the interference was based on sufficient reasons which rendered it 'necessary', regard must be had to the public-interest aspect of the case.'[46] This it found to be high, referring to the 'legitimate public concern' with the subject matter,[47] and noting that it was in actuality 'of evident concern to the local, national and international public'.[48]

The Court's strong declaration that animal welfare and the exposure of cruelty to animals is a legitimate matter of public interest is, in the author's opinion, more significant than the ultimate finding that article 10 had been violated in the circumstances of the case.[49] The Court has been loath to hold that restrictions on journalistic speech on matters of public interest can be justified except under circumstances of clear abuse, and has repeatedly held that the state's margin of appreciation will be very narrow where the freedom of the press is concerned.[50] That article 10 was

---

[41]ECHR, *Steel and Others* (n. 1), paras. 112-113; ECHR, *Drieman* (n. 1), 7-10.

[42]ECHR, *Bladet Tromsø* and *Stensaas* (n. 1).

[43]ECHR, *VgT Nos. I&II* (n. 1); ECHR, *Steel and Morris v. UK* (n. 1); ECHR, *PETA Deutschland v. Germany* (n. 1); ECHR, *ADI v. UK* (n. 1); and ECHR, *Tierbefreier e.V. v. Germany* (n. 1).

[44]ECHR, *Bladet Tromsø and Stensaas* (n. 1), paras. 6-38.

[45]Ibid., para. 59.

[46]Ibid., para. 62.

[47]Ibid., para. 64.

[48]Ibid., para. 63.

[49]Ibid., para. 73.

[50]See, for example, the statement by the Court in *Thorgeirson* that the press has a 'pre-eminent role [...] in a State governed by the rule of law': ECHR, *Thorgeirson v. Iceland* (n. 33), para. 63. See further, among others, *Sunday Times v. UK* (n. 33), 33; ECHR, *Lingens v. Austria* (n. 33), para. 13;

violated was an unsurprising conclusion, therefore, once the finding had been made that a debate on animal cruelty was a matter of public interest: it was the combination of public interest and journalistic speech which was important for the outcome. The same can be said of the subsequent article 10 cases concerning aspects of animal welfare that have come before the Court: for present purposes the ultimate decisions are of secondary importance, being determined primarily by factors not directly relevant to animal welfare. That is significant in itself, however, insofar as it demonstrates that the role given to animal welfare concerns in article 10 cases is a narrow one, limited to determination of the appropriate standard of review.

In the years following Bladet Tromsø *and Stensaas* the Court has ruled on the application of article 10 in relation to animal welfare in six cases, each of which concerned the legality of restrictions on communications by animal welfare groups either (as in *VgT Nos. I&II* and *ADI v. UK*) relating to the media available to them,[51] or (as in *Steel and Morris, PETA Deutschland* and *Tierbefreier e.V.*) the content of those communications.[52] In each case the Court reiterated its finding that animal welfare and animal rights are 'topics of general concern',[53] and 'questions of public interest'.[54] Accordingly, it has repeatedly noted that the margin of appreciation due to states in determining to what extent such communications can be restricted is narrow,[55] and has even gone so far as to suggest that the standard of protection appropriate to campaigning groups working on such matters is similar (although perhaps not identical) to that applicable to journalists.[56]

ECHR, *Oberschlick v. Austria*, Grand Chamber Judgment of 23 May 1991, Application No. 1162/85, para. 58; ECHR, *Observer and Guardian v. UK*, Grand Chamber Judgment of 26 November 1991, Application No. 13585/88, para. 59; ECHR, *Jersild v. Denmark* (n. 33), para. 31; ECHR, *Goodwin v. UK*, Grand Chamber Judgment of 27 March 1996, Application No. 17488/90, para. 39. These principles have been reaffirmed in the recent case of ECHR, *Satakunnan Markkinapörssi Oy and Satamedia Oy v. Finland*, Grand Chamber Judgment of 27 June 2017, Application No. 931/13, paras. 124-128.

[51]ECHR, *VgT No. I* (n. 1), paras. 8-23; *VgT No. II* (n. 1), paras. 12-27; ECHR, *ADI v. UK* (n. 1), paras. 8-33.

[52]ECHR, *Steel and Morris v. UK* (n. 1), paras. 8-36; ECHR, *PETA Deutschland v. Germany* (n. 1), paras. 6-19; ECHR, *Tierbefreier e.V. v. Germany* (n. 1), paras. 5-21.

[53]ECHR, *Steel and Morris v. UK* (n. 1), para. 88.

[54]ECHR, *PETA Deutschland v. Germany* (n. 1), para. 47. See also ECHR, *VgT No.II* (n. 1), para. 92.

[55]In *VgT No. I* the Court noted that 'in the present case the extent of the margin of appreciation is reduced, since what is at state is not a given individual's purely "commercial" interests, but [their] participation in a debate affecting the general interest': ECHR, *VgT No. I* (n. 1), para. 71. See also ECHR, *ADI v. UK* (n. 1), para. 104.

[56]ECHR, *Steel and Morris v. UK* (n. 1), para. 89, [references omitted].

4   Obstacles and Opportunities

Nevertheless, it is abundantly clear from the Courts' reasoning in these and other cases that what is protected is *the interest that humans may feel in the welfare and suffering of animals*, and not the welfare of animals as an end in itself. The distinction is illustrated particularly clearly in *PETA Deutschland v. Germany*. That case concerned an advertising campaign which juxtaposed images of mass farming methods with Nazi-era concentration camps, together with text which claimed similarities between the treatment of holocaust victims and treatment of animals in the modern meat industry. One such caption ran 'where animals are concerned everyone becomes a Nazi' ('[w]o es um Tiere geht, wird jeder zum Nazi').[57] An injunction was requested by Jewish community leaders, and granted by the domestic Courts on the basis that 'the debasement of concentration camp victims was [...] exploited in order to militate for better accommodation of laying hens and other animals.'[58] Though the courts at all levels noted that the campaign was not malicious in the sense that PETA did not intend to minimise the suffering of holocaust victims nor to violate their human dignity, they nevertheless concluded that the comparison was 'banalising', and that 'the Basic Law drew a clear distinction between human life and dignity on the one side and the interests of animal protection on the other'. The injunction was therefore justified on the basis that the 'content of the campaign affected the plaintiffs' personality rights.'[59]

The ECtHR unanimously agreed. Though its judgment noted the particular context of Jews living in Germany and pointed out 'that courts in other jurisdictions might address similar matters in a different way',[60] it accepted that the decision of the domestic courts was reasonable. In coming to that finding it agreed that the posters did not 'aim to debase the depicted concentration camp inmates, as the pictures merely implied that the suffering inflicted on the depicted humans and animals was equal.'[61] Nevertheless, and revealingly, it characterised the treatment of the concentration camp victims as 'instrumentalisation' in the 'interests of animal protection'.[62] This theme was taken up—and taken further—by Judge Zupančič in a concurring opinion joined by Judge Spielmann. There they asked

---

[57]ECHR, *PETA Deutschland v. Germany* (n. 1), para. 7.

[58]Ibid., para. 11.

[59]Ibid., paras. 17-18; see also Landgericht Berlin, Judgment of 18 March 2004, 27 O 207/04; Landgericht Berlin 22 April 2004, 27 O 207/04; Kammergericht, Judgment of 30 July 2004, 9 U 118/04; Kammergericht, Judgment of 27 August 2004, 9 U 118/04; Landesgericht Berlin, Judgment of 2 December 2004, 27 O 676/04; Kammergericht, Judgment of 25 November 2005, 9 U 15/05; Bundesverfassungsgericht, Judgment of 20 February 2009, 1 BvR 2266/04, 1 BvR 2620/05.

[60]ECHR, *PETA Deutschland v. Germany* (n. 1), para. 49.

[61]Ibid., para. 48.

[62]Ibid.

> [W]hether reasonable [people] could indeed or could not differ on the utterly disgraceful and unacceptable comparison between pigs on the one hand and the inmates of Auschwitz or some other concentration camp, on the other hand.[63]

> [W]hen human beings in their utter suffering and indignity are, as here, compared to hens and pigs for the lesser purpose of protecting otherwise legitimate advancement of animal rights, we are no longer in the position to maintain that the human beings seen in these pictures are treated as an end in themselves. [. . .] If their image is so instrumentalised, little is left of their human dignity[.][64]

Though these statements were seemingly too strongly put for the majority of the Court, the underlying reasoning appears to be the same. Certainly, Judge Zupančič is correct in pointing out that using the picture of the concentration camp victims simply as a comparator does not accord with the Kantian imperative that individuals be treated as ends in themselves, but it is worth pausing to consider whether a comparison of holocaust victims with—for example—victims of modern-day international crimes would have attracted the same condemnation. That would be no less of an instrumentalisation, but it seems clear that for Judges Zupančič and Spielmann (even if not for the domestic courts) the context of the comparison was a significant part of the harm the campaign committed. It may be that it was not instrumentalisation *per se* that was objectionable, but rather instrumentalisation of the human in service of the animal.

Nor is that conclusion—or its counterpart that the animal may be instrumentalised in service of the human—a surprising position for the Court to reach. It is almost unnecessary to say, for example, that the Court has no objection in principle to the use of animals as game,[65] for medical experimentation,[66] or for food.[67] In some circumstances animal welfare concerns take second place to facilitation of religious practice.[68] Given a direct conflict between animal life and human safety and wellbeing the Court unsurprisingly privileges human safety.[69] The Court should

---

[63]Concurring Opinion of Judge Zupančič, Joined by Judge Spielmann, ECHR, *PETA Deutschland v. Germany* (n. 1), 16-18, at para. 5.

[64]Ibid., paras. 14-15. A similar argument in the academic sphere is made by Leslie Pickering Francis and Richard Norman, who argue that the term "animal liberation" 'has the effect of trivializing [. . .] real liberation movements, putting them on a level with what cannot but appear as a bizarre exaggeration': Francis/Norman, 'Some Animals are More Equal than Others' 1978, 527. Kymlicka and Donaldson respond powerfully to such arguments: Kymlicka/Donaldson, 'Animal Rights, Multiculturalism, and the Left' 2014, 116-135. A historical analysis of the human/animal dichotomy is given by Anna Becker in her contribution to this volume.

[65]See ECHR, *Chassagnou v. France* (n. 1), and the cases that followed it.

[66]ECHR, *Tierbefreier e.V. v. Germany* (n. 1).

[67]ECHR *Cha'are Shalom* (n. 1); ECHR, *Jakóbski v. Poland* (n. 1).

[68]ECHR *Cha'are Shalom* (n. 1). Importantly, though, the Court held here that the state retained a margin of appreciation to decide on what basis permits to slaughter animals in accordance with religious requirements (in this case the strict requirements to qualify as glatt kosher) would be granted in order to, among other things, enable it to protect public health and animal welfare (paras. 76-77, 84). Provided that meat prepared according to the requirements of one's religion is available, article 9 does not extend to a right to slaughter one's meat oneself (paras. 80-82).

[69]ECHR, *Stoicescu v. Romania* (n. 1).

not be criticised for these positions: to adopt the contrary finding on any of these points would take the Court radically beyond the understanding of the rights involved prevalent in the Council of Europe states, and thus well beyond its remit to 'interpret[the Convention] in the light of present-day conditions and of the ideas prevailing in democratic States'.[70] The Court generally cites *Tyrer v. UK* as the source of this principle, which casts the Convention as a 'living instrument'.[71] Although this means that standards of protection may develop,[72] the Court has also explicitly arrived at the corollary conclusion that it should not go beyond these evolving standards.[73] It is therefore not at liberty to find that hunting, eating or experimenting on animals is improper even if it were inclined to do so: every Council of Europe state accepts these practices within certain limits, and nor is there a consensus even among animal activists and scholars on their (im)propriety.[74]

Although the obligation—both precedential and of prudence—not to stray beyond the understanding of the Convention rights among the states forecloses certain radical steps in using the ECHR to protect animal welfare, the 'living instrument' formulation also offers the promise that future developments may be incorporated into the Convention's protections. Judge Pinto de Albuquerque's separate opinion in *Hermann v. Germany* in 2012 gives an indication of the mechanism through which this could take place.[75] Animals, Pinto de Albuquerque argued, are protected under the ECHR in two ways. First, they may be property within the definition of article 1 of Protocol 1.[76] More importantly, they may be protected 'as beings in themselves [. . .] as part of a healthy, balanced and sustainable environment',[77] under the umbrella of the article 8 obligation to 'avoid acts and activities that could have detrimental consequences for public health and the environment'.[78] Pinto de Albuquerque finds "'clear and uncontested evidence of a continuing international trend" in favour of the protection of animal life and welfare

---

[70]ECHR, *Khamtokhu and Aksenchik v. Russia*, Grand Chamber Judgment of 24 January 2017, Application Nos. 60367/08 and 961/11, para. 73 [references omitted].

[71]ECHR, *Tyrer v. UK*, Chamber Judgment of 25 April 1978, Application No. 5856/72, para. 31.

[72]ECHR, *Khamtokhu* and Aksenchik (n. 70), para. 73; see also ECHR, *Selmoui v. France*, Grand Chamber Judgment of 28 July 1999, Application No. 25803/94, para. 101.

[73]ECHR, *Khamtokhu and Aksenchik* (n. 70), para. 74

[74]For different perspectives on these questions see Cochrane, *Animal Rights without Liberation: Applied Ethics and Human Obligations* 2012; Taylor, 'Whiter Rights? Animal Rights and the Rise of New Welfarism' 1999, 27-41; Harrop, 'Climate Change, Conservation and the Place for Wild Animal Welfare in International Law' 2011, 441-462. Outside academia, compare the remit of the Animal Welfare Council (http://www.animalwelfarecouncil.org/?page_id=9), with PETA (https://www.peta.org/about-peta/faq/what-is-the-difference-between-animal-rights-and-animal-welfare/).

[75]Partly Concurring and Partly Dissenting Opinion of Judge Pinto de Albuquerque, ECHR, *Herrmann v. Germany* (n. 1), 32-49. Similar themes were also discussed in the earlier Partly Dissenting Opinion of Judge Zagrebelsky, ECHR, *Kyrtatos v. Greece* (n. 1), 14-15.

[76]Ibid., 32.

[77]Ibid.

[78]Ibid., 33 [references omitted].

[which] is reflected in the application of the Convention.'[79] He argues that the Court should reject both the 'commodification' of animals and extensive conceptions of human-like animal personality, instead embracing a 'qualified speciesism which builds upon a responsible anthropocentrism.'[80] He concludes that recognising the moral differences between humans and animals 'does not prevent us from acknowledging the [. . .] existence of basic comparable interests between humans and other animals and therefore the need to safeguard certain 'animal rights', metaphorically speaking, in a similar way to human rights.'[81] The mechanism through which this should be achieved is not the grant of legal personality to animals to raise claims before the Court (nor upon human 'representatives' to do so),[82] but rather through the obligation of states to realise the human right to a healthy environment.[83]

While there is much here that is attractive, there remain problems with the application of the approach Pinto de Albuquerque proposes, and flaws in the approach itself. To begin with application, it is increasingly accepted that a healthy environment is an aspect of human rights.[84] As yet, however, it is unclear whether the ECHR has the potential adequately to integrate this idea into its provisions. Prima facie, environmental harms are more closely connected to the protection of social and economic rights than the primarily civil and political rights of the ECHR. The disconnect is clear in *Kyrtatos v. Greece*, in which the Court was asked to decide that the illegal destruction of a wetland habitat next to the applicants' house was a violation of article 8. The Court chose not to do so, holding that the applicants had not demonstrated that the effect of the environmental degradation on them 'directly affect[ed] their own rights under article 8'.[85] The Court reached that conclusion by six votes to one, with Judge Zagrebelsky the only dissenter. It can be speculated then, that although it would be possible for environmental degradation to have sufficiently negative effects to amount to a breach of article 8, such a finding is likely to be made only where there is a measureable negative effect on individuals' health or some other equally weighty aspect of their lives. By contrast, and despite that it materially affected their quality of life, 'the Court

---

[79]Ibid., 36, citing ECHR, *Goodwin v. UK* (n. 50), para. 85.

[80]Opinion of Pinto de Albuquerque (n. 75), 37 [emphasis and references omitted].

[81]Ibid., 37 [references omitted].

[82]The ability of animals to appear as "persons" before the courts is discussed below, at note 89.

[83]Opinion of Pinto de Albuquerque (n. 75), 38.

[84]See, for example, the recent framework principles prepared by John Knox in his capacity as special rapporteur: Human Rights Council, 'Report of the Special Rapporteur on the Issue of Human Rights Obligations relating to the Enjoyment of a Safe, Clean, Healthy and Sustainable Environment', 24 January 2018, UN Document No. A/HRC/37/59. Knox suggests two parallel provisions as his first and second framework principles, that '[s]tates should ensure a safe, clean, healthy and sustainable environment in order to respect, protect and fulfil human rights', and that '[s]tates should respect, protect and fulfil human rights in order to ensure a safe, clean, healthy and sustainable environment' (page 7).

[85]ECHR, *Kyrtatos v. Greece* (n. 1), para. 53.

[did not] accept that the interference with the conditions of animal life in the swamp constitute[d] an attack on the private or family life of the applicants.'[86] If the precedent set in *Kyrtatos* stands, then, a harm to animal life and the wider environment will have to produce very substantial negative impacts on individuals before it will be possible to assimilate these harms under article 8.[87]

Yet there are potential problems, too, with the idea of responsible anthropocentricism as a theoretical lens through which to interpret the ECHR in ways conducive to the protection of animal welfare, in that it remains—obviously— anthropocentric.[88] Of course, one could hardly expect the ECtHR to move to a position *beyond* 'responsible anthropocentricism' without alteration of the Convention or a substantial leap in its interpretation. Such an interpretive move would, in theory, be possible: indeed, there is nothing in the text of the convention that would prevent it from being extended to apply to (some) animals. Despite its title (Convention for the Protection of *Human* Rights; Convention de sauvegarde des droits de *l'homme*), the personal scope of the Convention as defined by its first article does not refer to 'humans' but rather to 'everyone' and 'toute personne', both terms which seem amenable to a legal rather than scientific definition. Nevertheless, it remains to be seen whether the Court is able to go this far,[89] and that uncertainty serves to make

---

[86]Ibid.

[87]This conclusion is broadly supported by Natalia Kobylarz's study of the Court's case-law on wider aspects of environmental protection. While the Court has been able to provide relief under the ECHR in a number of environmental damage scenarios, it remains necessary to show an immediate link to a concrete harm. See Natalia Kobylarz, 'The European Court of Human Rights: An Underrated Forum for Environmental Litigation', in European Environmental Law Forum, Sustainable Management of Natural Resources – Legal Approaches and Instruments (forthcoming), available at: https://ssrn.com/abstract=3178983, accessed 30/05/2018.

[88]On this idea see Redgwell, 'Life, The Universe And Everything' 1996, 71-87, esp. 75-79; Shelton, 'Environmental Rights' 2001, 190; Bulto, 'The Environment and Human Rights' 2014, 1015-1030; de Lucia, 'Beyond Anthropocentrism and Egocentrism', 2017, esp. 184-188.

[89]Note, for example, the Court's decision to deny jurisdiction ratione personae over the application in *Stibbe v. Austria*, a case brought by an animal rights activist on behalf of a chimpanzee known as Matthis Pan. Stibbe sought to be appointed the legal guardian of Matthis Pan, but her application was denied by the Austrian courts on the basis that only humans can have guardians. Her appeal to the ECtHR was declared inadmissible on the basis that '[t]he applicant cannot [. . .] claim to have herself been a victim of the violation in accordance with article 34 of the Convention. The complaint is therefore not in accordance with the personal scope of the Convention under article 35 paragraph 3': Letter from A. Wampach, Deputy Registrar for the First Section, in the matter of ECHR, *Stibbe v. Austria*, 22 January 2010, Reference No. ECHR-LGer11.0R(CD8); IF/IW/tpe; Application No. 26188/08 [my translation]. Jurisdiction *ratione personae*, clearly, will be an obstacle to cases of this kind being heard before the ECtHR. This may be contrasted to the now-famous *Orangutána Sandra* decision before the courts of Argentina, in which it was decided animals may be the subject of rights: 'Based on a dynamic rather than a static legal interpretation, it is necessary to accord the animal the status of a rights-holder. Non-human subjects (animals) are bearers of rights, and therefore their protection is required within the corresponding jurisdiction': Camera Federal de Casación Penal, *Orangutána Sandra*, Judgment of 18 December 2014, LEX No. CCC 68831/2014/ CFC1, para. 2. [I thank Dr Pedro Villarreal for his assistance interpreting the judgment and preparing this translation.] Similar decisions were handed down in 2016 in another Argentinian

the anthropocentrism problem and its study more urgent. If anthropocentricism is a barrier to the formulation of meaningful principles to undergird animal welfare (let alone animal rights), then one must necessarily conclude that the human rights framework cannot contribute to the development of global animal law.[90] That question has been discussed elsewhere (and is taken up in several of the other contributions to this volume), and is too large and complex adequately to be discussed here.[91] However, in the present author's opinion, this proposition is not correct. On the contrary, human rights law can meaningfully contribute to the development of global animal law. Though it may be that global animal law will eventually need to separate itself from human rights law if it is to realise its potential, in its early stages of development there are numerous opportunities for synergistic interactions with frameworks such as the ECHR.

This is the argument forcefully and convincingly made by Connor Gearty in the wider context of environmental protection.[92] Gearty begins by acknowledging that environmental concerns (and, for our purposes, animal welfare and rights) do not sit easily alongside the human rights framework's proud anthropocentricism:

> The subject of human rights is, as it declares for all to see in the way that it describes itself, a field that is concerned not only with humans but also with the rights that flow from being human, rather than from being anything else[.][93]

Human rights law exemplifies and makes explicit a sin Anne Peters identifies more generally, that 'the law as it stands mirrors and reifies a human-animal divide'.[94] Yet

---

case (Tercer Juzgado de Garantías de Mendoza, *Chimpanzee 'Cecilia'*, Judgment of 3 November 2016, No. P-72.254/15), and by the Colombian Supreme Court in 2017, granting habeas corpus in favour of a spectacled bear: Corte Suprema de Justicia de Colombia, Judgment of 26 July 2017, AHC4806–2017, Radicación no. l7001–22–13–000–2017–00468–02. In the common law world such cases have to date been raised only in the USA, and as yet without great success. In the most recent development (at time of writing), application to appeal to the New York Court of Appeals was denied on 5 April 2018 in joined cases submitted on behalf of two chimpanzees, in which a writ of habeas corpus was denied at first instance: State of New York Court of Appeals, *In re the Nonhuman Rights Project, inc., on behalf of Tommv v. Patrick C. Lavery* and *In re the Nonhuman Rights Project, inc., on behalf of Kiko v. Carmen Presti et al.*, Judgment of 8 May 2018, unreported, Motion No. 2018-268.

[90]And there are many who argue that it *should* not. See, for example, Elder, 'Legal Rights for Nature – The Wrong Answer to the Right Question' 1984, 285-295; Livingston, 'Rightness or Rights?' 1984, 309-321; Machan, 'Do Animals Have Rights?' 1991, 163-173; Merrills, 'Environmental Rights' 2007, 672.

[91]See, in particular, Peters, 'Liberté, Égalité, Animalité: Human-Animal Comparisons in Law' 2016b, 39-44 et seq.; Gearty, 'Do Human Rights Help or Hinder Environmental Protection' 2010, 7-22; and further Plass, 'Exploring Animal Rights as an Imperative for Human Welfare' 2010, 403-430; Keim/Sosnowski, 'Human Rights v Animal Rights: Mutually Exclusive or Complementary Causes' 2012, 78-83. An intriguing (but, in the author's view, ultimately ill-directed) inversion of this debate is Shikubu, 'Work like a Dog' 2014, 44-65.

[92]Gearty, 'Human Rights and Environmental Protection' 2010.

[93]Ibid., 7. [References omitted]. A similar argument is made by Knox, 'Climate Ethics and Human Rights' 2014, 22-34; but compare the problematisation of this aspect of human rights discourse in Blouin Genest/Paquerot, 'Environmental Human Rights as a Battlefield' 2016, 132-154.

[94]Peters, 'Liberté, Égalité, Animalité' 2016a, 26.

Gearty argues that human rights has the potential to support environmental protection both through the protection of environmental activism ('protecting the messenger'),[95] and by offering a vocabulary of empowerment that activists can use.

> It speaks meaningfully across the whole spectrum of a community, from the weak across to the powerful, deploying the convictions of the latter—rooted in the battles of the past—to force recognition of the need for similar struggles today. [...] This chameleonism is often a source of frustration for sure, but it is what gives the idea of human rights the power that it undeniably enjoys in the world today.[96]

Though human rights are intrinsically anthropocentric, the human rights project is a legally-embedded socio-linguistic mobilisation of empathy for the other.[97] Using the language of human rights carries with it the historical experience of the manifold struggles for justice that have been fought under its banner. Embedded in the framework are the memories of many claims once bitterly contested as radical oppositions to an entrenched power-structure which have succeeded in breaking into the mainstream consciousness, have overturned centuries of social practice, or have been codified as a minimum standard of positive morality in international declarations and conventions. The language, experience, and historical legitimacy-claim of human rights can be powerful tools in the campaign for animal (and wider environmental) rights, notwithstanding the inevitable friction between zoo- and anthropos-centrism.

5 Final Thoughts

Although that friction is more pronounced (and the radical discourse more constrained) within human rights viewed as a legal framework rather than a socio-political project, nevertheless many of the same arguments hold true. There are barriers to the direct treatment of animal concerns by human rights fora as a result of personal and material limitations on their scope of jurisdiction, but the case-law of the ECtHR demonstrates that there remain opportunities to bring animal concerns under the umbrella of human welfare. This does, it is true, raise moral questions, in particular the "speaking for the other' problem', as Catharine MacKinnon has

---

[95]Gearty, 'Human Rights and Environmental Protection' 2010, 15-18.

[96]Ibid., 21; for a similar argument grounded in the concept of dignity see Kotzmann/Seery, 'Dignity in International Human Rights Law' 2017, 1-41.

[97]Gearty, 'Human Rights and Environmental Protection' 2010, 22. The significance of empathy is also persuasively emphasised by Peters, who notes not only the transformative power of empathy on discourses and societies (39-42), but also the potential for definitions to structure empathic reactions. She begins by recalling the hideous nineteenth and twentieth century practice of displaying people of non-European origin as zoo exhibits, and notes that '[t]he "primitives" were relegated to the animal side of an imagined boundary': Peters, 'Liberté, Égalité, Animalité' 2016a, 25-26; see also Peters, 'Introduction: Animal Law – A Paradigm Change' 2015, 17-18. For an examination of empathy as a basis for distinctively human rights see Robinson, 'Biological Foundations of Human Rights' 2013, 54-81.

pointed out.[98] Animal law remains human law, and it aspires towards a human interpretation of what 'animal welfare' looks like. Yet though the interpretative divide is deeper, Peters is clearly correct to ask where the differences lie between speaking for animals and speaking for humans who lack legal capacity (Peters' example is children).[99] Arguably in the case of animals the situation is more problematic: where we raise children's concerns before Courts we do so for the benefit of the children involved, while animal rights at present flow from human rights only as a corollary of human concerns. The former is a case of speaking for, with all the moral difficulties that flow from that; the latter is an example of instrumentalisation. Yet there is also a zone of confluence,[100] in which human and animal wellbeing and rights coincide insofar as it can be demonstrated that protecting the one benefits the other.[101] Peters uses the phrase 'liberté, égalité, animalité' as 'a reminder that humans need legal protection not least on account of their animal nature, their physical vulnerability and their "nakedness", which they share with all other animals.'[102] It is indeed a salutary reminder that the human/animal divide is bridged in many respects, including the 'vital interests' of both groups.[103] Articulating those confluences within the language of the ECHR and other human rights frameworks has the potential to catalyse the development of animal welfare as a sub-genre of the international human rights story, as well as to provide norms, ideas and impetuses which will cross into other jurisdictions and disciplines, and scholars should now take up this task. It is in these interactions that global animal law is growing and will continue to grow,[104] and this brief

---

[98]MacKinnon, 'Of Mice and Men: A Feminist Fragment on Animal Rights' 2004, 270.

[99]Peters, 'Liberté, Égalité, Animalité' 2016a, 48.

[100]This idea is similar to Bulto's *substantive regime complementarity*: Bulto, 'Environment' 2014, 1025-1028

[101]An example of such an approach in practice can be seen in the Court's decision on admissibility in *Friend and Others v UK* (n. 1). In that case, a challenge to the UK ban on hunting wild mammals with dogs, the Court first ruled that the Convention articles claimed by the applicants were not engaged, before noting (in particular in relation to article 11) that 'the measures served the legitimate aim of (...) "the protection of ... morals", in the sense that they were designed to eliminate the hunting and killing of animals for sport in a manner which the legislature judged to cause suffering and to be morally and ethically objectionable' (at 18). The Court thus found that had the convention rights been engaged, the limitation would nevertheless have fallen within the State's margin of appreciation. Though at best indicative, as no full examination was undertaken, the admissibility decision shows one way in which the interests of animals can condition human rights—in this case as a limitation, elsewhere through a zone of confluence approach.

[102]Peters, 'Liberté, Égalité, Animalité' 2016a, 53; citing Saskia Stucki, 'Sind die Menschenrechte in Zukunft noch Menschen-Rechte?', Völkerrechtsblog, 13 May 2014, available at: http://voelkerrechtsblog.com/category/sind-die-menschenrechte-in-zukunft-noch-menschen-rechte/.

[103]This idea I take from Mark Rowlands, *Animals Like Us* (London: Verso 2002), 125-136 et seq. Rowlands uses the term to refer to the interest all animals have in remaining alive, as well as the basic goods that enable them to do so. He argues that the non-vital interests of any (human or non-human) animal should not outweigh the vital interests of any other.

[104]Anne Peters, 'Global Animal Law: What it is and why we need it', *Transnational Environmental Law* 5(1) (2016b), 9-23, 20.

examination of the ECtHR suggests that human rights law has a meaningful contribution to make to that process.

References

Blouin Genest, G., & Paquerot, S. (2016). Environmental human rights as a Battlefield: A grammar of political confrontation. *Journal of Human Rights and the Environment, 7*, 132–154.
Bulto, T. S. (2014). The environment and human rights. In A. Mihr & M. Gibney (Eds.), *The SAGE handbook of human rights* (Vol. II, pp. 1015–1030). London: Sage Publishing.
Çoban, A. R. (2004). *Protection of property rights within the European Convention on Human Rights*. Aldershot: Ashgate.
Cochrane, A. (2012). *Animal rights without liberation: Applied ethics and human obligations*. New York: Columbia University Press.
de Lucia, V. (2017). Beyond anthropocentricism and egocentrism: A biopolitical reading of environmental law. *Journal of Human Rights and the Environment, 8*(2), 181–202.
Elder, P. S. (1984). Legal rights for nature – the wrong answer to the right question. *Osgoode Hall Law Journal, 22*(2), 285–295.
Fenwick, H., Phillipson, G., & Williams, A. (2017). *Texts, cases and materials on public law and human rights* (4th ed.). Abingdon: Routledge.
Gearty, C. (2010). Do human rights help or hinder environmental protection. *Journal of Human Rights and the Environment, 1*, 7–22.
Harrop, S. (2011). Climate change, conservation and the place for wild animal welfare in international law. *Journal of Environmental Law, 23*(3), 441–462.
Keim, S., & Sosnowski, J. (2012). Human rights v animal rights: Mutually exclusive or complementary causes. *Australian Animal Protection Law Journal, 8*, 78–83.
Knox, J. (2014). Climate ethics and human rights. *Journal of Human Rights and the Environment, 5* (Special Issue), 22–34.
Kobylarz, N. (forthcoming). The European Court of Human Rights: An underrated forum for environmental litigation. In *European Environmental Law Forum, Sustainable Management of Natural Resources – Legal Approaches and Instruments*. Available at: https://ssrn.com/abstract=3178983
Kotzmann, J., & Seery, C. (2017). Dignity in international human rights law: Potential applicability in relation to international recognition of animal rights. *Michigan State International Law Review, 26*(1), 1–41.
Kymlicka, W., & Donaldson, S. (2014). Animal rights, multiculturalism, and the left. *Journal of Social Philosophy, 45*(1), 116–135.
Livingston, J. (1984). Rightness or rights? *Osgoode Hall Law Journal, 22*(2), 309–321.
Machan, T. (1991). Do animals have rights? *Public Affairs Quarterly, 5*(2), 163–173.
MacKinnon, C. (2004). Of mice and men: A feminist fragment on animal rights. In C. Sunstein & M. Nussbaum (Eds.), *Animal rights: Current debates and new directions* (pp. 263–276). Oxford: OUP.
Merrills, J. (2007). Environmental rights. In D. Bodansky, J. Brunnée, & E. Hey (Eds.), *The Oxford handbook of international environmental law* (pp. 663–680). Oxford: OUP.
Peters, A. (2015). Introduction: Animal law – a paradigm change. In A. Peters, S. Stucki, & L. Boscardin (Eds.), *Animal law: Reform or revolution* (pp. 15–32). Zürich: Schulthess.
Peters, A. (2016a). Liberté, Égalité, Animalité: Human-Animal Comparisons in Law. *Transnational Environmental Law, 5*, 25–53.
Peters, A. (2016b). Global animal law: What it is and why we need it. *Transnational Environmental Law, 5*(1), 9–23.

Plass, S. A. (2010). Exploring animal rights as an imperative for human welfare. *West Virginia Law Review, 112*, 403–430.

Redgwell, C. (1996). Life, the universe and everything: A critique of anthropocentric rights. In A. Boyle & M. Anderson (Eds.), *Human rights approaches to environmental protection* (pp. 71–87). Oxford: Clarendon Press.

Robinson, C. (2013). Biological foundations of human rights. In D. Shelton (Ed.), *The Oxford handbook of international human rights law* (pp. 54–81). Oxford: OUP.

Rowlands, M. (2002). *Animals like us*. London: Verso.

Shelton, D. (2001). Environmental Rights. In P. Alston (Ed.), *Peoples' rights* (pp. 185–258). Oxford: OUP.

Shikubu, M. (2014). Work like a dog: Expanding animal cruelty statutes to gain human rights for migrant farmworkers in the US. *National Lawyers Guild Review, 71*, 44–65.

Stucki, S. (2014, May 13). Sind die Menschenrechte in Zukunft noch Menschen-Rechte?. *Völkerrechtsblog*. Available at http://voelkerrechtsblog.com/category/sind-die-menschenrechte-in-zukunft-noch-menschen-rechte/

Taylor, N. (1999). Whiter rights? Animal rights and the rise of New Welfarism. *Animal Issues, 3*(1), 27–41.

Tom Sparks is a senior research fellow at the Max Planck Institute for Comparative Public Law and International Law in Heidelberg, where he works in the research group of Professor Anne Peters. His research interests focus on public international law, international environmental law, the humanisation of law, and legal theory.

Chapter 14
Challenges Regarding the Protection of Animals During Warfare

Jérôme de Hemptinne

Abstract The chapter turns to the treatment of animals in one of the two classical divisions of international law, the laws of war, examining the protection of animals during hostilities. De Hemptinne explains that international humanitarian law (IHL) does not contain explicit rules to mitigate the suffering of animals in armed conflict. However, the overall evolution of law's approach to animals, notably its recognition of them as sentient beings, appears to allow for a progressive interpretation of IHL so as to constrain acts of violence against animals in war. The rules on the protection of civilian objects and on the environment, the proportionality principle, or the options for declaring demilitarized zones could all be activated to this end.

1 Introduction

In times of war, the first instinct is to relieve the suffering of human beings. Environmental and animal interests are always pushed into the background. However, warfare strongly affects natural resources, including animals, which makes animal issues a matter of great concern. Habitat destruction and the resulting disappearance of animals often threaten the survival of populations affected by hostilities. Furthermore, over the last 50 years, certain species have been vanishing at a rapid rate because of wars, often with disastrous effects on the food chain and on the ecological balance. Indeed, during this period, 80% of armed conflicts have taken place in countries—such as Afghanistan, Burundi, Central African Republic, Congo,

Revised version of the original published article "The Protection of Animals during Warfare" by Jérôme de Hemptinne, American Journal of International Law Unbound, Volume 111, 2017, pp. 272–276. The original article was published as an Open Access article, distributed under the terms of the Creative Commons Attribution licence (http://creativecommons.org/licenses/by/4.0/).

J. de Hemptinne (✉)
University of Geneva, Geneva, Switzerland

A. Peters (ed.), *Studies in Global Animal Law*, Beiträge zum ausländischen öffentlichen Recht und Völkerrecht 290,
https://doi.org/10.1007/978-3-662-60756-5_14

Kenya, Rwanda, Uganda or Vietnam—that contain areas of high global species diversities.[1] Belligerents even take advantage of the chaotic circumstances of war in order to poach protected species and to engage in the trafficking of expensive animal products. While generating billions of dollars each year—which are, in part, invested in warfare and the acquisition of weapons—such poaching and trafficking allows armed groups to grow and to reinforce their authority over disputed territory. This fuels a cycle of violence and ultimately threatens peace and security in these areas. States have also trained, and continue to train, certain animals—principally marine mammals such as bottlenose dolphins and California sea lions—to perform military tasks, like ship and harbour protection, or mine detection and clearance. Millions of horses, mules, donkeys, camels, dogs and birds are obliged to serve on various fronts (transport, logistics, or communications) and become particularly vulnerable targets.

However, being deeply anthropocentric, international humanitarian law (IHL) largely ignores the protection of animals. That said, some general principles could potentially provide minimum safeguards to animals during armed conflict. Moreover, a progressive interpretation of these principles, in light of developments in the welfare and rights of animals in peacetime, could significantly reinforce this protection. After having discussed the reasons underlying the silence of IHL on the protection of animals (Sect. 2), this chapter will outline the particular challenges that this issue creates in the context of the distinction between international armed conflicts (IACs) and non-international armed conflicts (NIACs) (Sect. 3). Finally, it will consider the main difficulties stemming from the rules governing the conduct of hostilities (Sect. 4), and from the protective regime offered by the 1949 Geneva Conventions (GCs)[2] (Sect. 5).

2   The Silence of IHL

Despite the fact that animals are among those affected by armed conflicts, IHL does not directly deal with the question of their protection. As we will see below, animals are only indirectly addressed as civilian objects or as part of the natural environment. Three reasons might explain IHL's lack of interest in the welfare of animals. First, the main conventions regulating armed conflicts were adopted at a time when legal entitlements for animals did not attract significant attention. Geared essentially towards the safeguarding of human beings, IHL was—and still remains today—an

---

[1]Hanson et al., 'Warfare in Biodiversity Hotspots' 2009, 578. See generally Daskin/Pringle, 'Warfare and wildlife declines in Africa's protected areas' 2018, 328-332.

[2]Geneva Convention (I) for the Amelioration of the Condition of the Wounded and Sick in Armed Forces in the Field, 12 August 1949, 75 UNTS 31. Geneva Convention (II) for the Amelioration of the Condition of Wounded, Sick and Shipwrecked Members of Armed Forces at Sea, 12 August 1949, 75 UNTS 85. Geneva Convention (III) relative to the Treatment of Prisoners of War, 12 August 1949, 75 UNTS 135. Geneva Convention (IV) relative to the Protection of Civilian Persons in Time of War, 12 August 1949, 75 UNTS 287.

overwhelmingly anthropocentric body of law 'displaying a strong utilitarian fla-
vour.'[3] Indeed, even the few IHL provisions restricting attacks against the natural
environment were designed with a view to preserving the interests of human beings
instead of the environment per se.[4] The Christian roots of IHL might also explain
such an anthropocentric attitude and the difficulty for this body of law to concep-
tualise the protection of the environment independently of human interests. Second, on a more pragmatic level, it is often argued that alleviating the extreme
suffering of human beings during hostilities would be impaired should animal
protection be put at the forefront. Devoting time, energy and money to protect
animals would prevent using these resources for worthier human causes. Lastly,
we should not underestimate the fact that, in peacetime, while acts of violence
against human beings are, in principle, forbidden, the slaughtering of animals—in
conformity with certain methods and procedures—is widely accepted and prac-
tised throughout the world. In this context, it might appear paradoxical that, in
situations of armed conflict, where acts of violence against certain individuals,
mainly combatants, are admitted and expected, acts of violence against animals
might be strictly constrained.

Let us briefly respond to these three arguments. First, IHL cannot ignore the
evolution of the status and protection that animals have acquired in many jurisdictions around the world. In the same way that the strengthening of human rights and
the increased awareness of environmental challenges after the Second World War
have impacted on the development of IHL, the increasing concern for animal welfare
during the last decades should also progressively impose limits on belligerents'
actions. This is even more so when we consider that, as recalled in the introduction,
wars have disastrous effects on wildlife and, in particular, on protected species
whose survival is directly threatened by hostilities. Second, contrary to what is
often claimed, safeguarding animals does not necessarily interfere, or run contrary
to, the protection of human beings. The two can often run in parallel without
impacting negatively on each other. In any case, we will see that certain IHL
principles—such as the principle of proportionality—are flexible enough to guarantee that human values prevail over animal interests when one is forced to choose
between the two. Third, the apparent 'paradox of violence' outlined above must be
evaluated in light of the following factors. Due to the increased importance of animal
welfare, acts of cruelty against animals are now widely restricted and sanctioned in
peacetime. Conversely, acts of violence against animals during warfare could, in
theory, be legally committed when animals become military objectives. Moreover,
the forms of violence allowed in peacetime are of a fundamentally different nature
from those authorised during armed conflicts. Indeed, the former aims at satisfying
human needs (for instance, food production or medical, pharmaceutical and chemical testing). In contrast, the latter is, in principle, dictated by military considerations.
This delineation between different types of violence is, however, not that

---

[3]Schmitt, 'Green War' 1997, 6.
[4]Ibid., 69.

straightforward: the slaughtering of animals for human uses also occurs in wartime, while certain forms of violence, which are usually committed in the chaotic circumstances of war—such as poaching and trafficking of species—can also take place during peacetime.

3   The Distinction Between IACs and NIACs

It is well know that the laws governing IACs—that is hostilities which oppose the armed forces of a state to another state—are much more developed than the laws regulating NIACs—that is hostilities of a certain intensity opposing armed groups to state armed forces or armed groups between themselves. In IACs, animals could, in theory, benefit from the protection offered by advanced IHL rules contained in the GCs and in the 1977 Additional Protocol I (AP I).[5] Animals should also benefit from the fact that states possess developed institutional machinery, well-organised military forces, and sophisticated technologies that should improve the respect of these rules. The situation is, however, different in NIACs owing to legal and practical difficulties.

From a legal perspective, it should be recalled that only minimal conventional rules apply to NIACs (i.e. common Article 3 to the GCs and Additional Protocol II (AP II)[6]). That said, the recent development of customary IHL has narrowed the differences in regulation between IACs and NIACs. For instance, it is now widely accepted that the general rules governing the protection of civilian objects—namely, the principles of military necessity/humanity, distinction and proportionality—apply to all types of armed conflicts. This is also true for the rules protecting enemy property from wanton destruction[7] and objects indispensable to the survival of the civilian population,[8] and for the rules prohibiting the use of certain weapons,[9] such as biological or chemical weapons. This evolution is crucial, because, as we will see in Sect. 3 below, all of these rules could be interpreted as providing minimum protection for animals, should they be treated as objects. In contrast, it is unclear whether the special protection accorded to the environment by Articles 35(3) and 55(1) of AP I—which encompasses wildlife[10]—applies in NIACs within the state

---

[5]Protocol Additional to the Geneva Conventions of 12 August 1949, and relating to the Protection of Victims of International Armed Conflicts (Protocol I), 8 June 1977, 1125 UNTS 3.

[6]Protocol Additional to the Geneva Conventions of 12 August 1949, and relating to the Protection of Victims of Non-International Armed Conflicts (Protocol II), 8 June 1977, 1125 UNTS 609.

[7]GC I, 1949 (n. 2), art. 50. GC II, 1949 (n. 2), art. 51. GC IV, 1949 (n. 2), art. 147.

[8]AP I, 1977 (n. 5), art. 54.

[9]AP I, 1977 (n. 5), art. 35(2) & art. 51(4).

[10]Roscini, 'Animals and the Law of Armed Conflict' 2017, 61.

where such a conflict is taking place.[11] This clearly reflects the reluctance of governments to accept that heavy constraints be imposed on them to conduct hostilities against rebels on their national territories. It could, however, be argued that the environment should always be safeguarded in NIACs when damages caused to it also affect other states. In this context, since wildlife is usually not confined to the territory of a single state and since its destruction generally affects the ecological balance on a wide scale, one could reasonably argue that the provisions protecting the environment contained in AP I should always protect wildlife, including within the territory of a state which experiences a NIAC.

From a practical perspective, armed groups often have limited abilities to conform to highly sophisticated norms, in particular those protecting animals. Indeed, assessing the legality of incidental damage caused to animals is beset with difficulties. It requires environmental knowledge that even well-equipped state armed forces often do not possess, particularly when considering that these damages tend to manifest themselves in the long run and that they are often the indirect result of the destruction of certain habitats. We will see below how complex it is to factor animal considerations into the application of the main IHL principles, in particular, when evaluating proportionality. Furthermore, the logic which underlies IHL— according to which the legitimate aim of belligerents is to weaken the military potential of the adversary[12]—will often be called into question when armed groups are motivated by other goals, such as poaching protected species and engaging in the trafficking of animal products. Thus, in the context of NIACs, ensuring the respect of complex IHL rules protecting the basic needs of animals will often be very difficult to achieve in practice.

4 The Conduct of Hostilities

Civilian objects are protected during the conduct of hostilities unless and for such time that they are military objectives.[13] Moreover, according to the principle of proportionality, collateral damage on civilian objects is permissible but only to the extent that it is not excessive in relation to the concrete and direct military advantage anticipated as the result of an attack.[14] These basic principles raise three main delicate questions in relation to animals. First, could animals be treated as civilian objects for the purpose of IHL? Second, if the first question is answered in the affirmative, how should one undertake the balancing exercise required by the principle of proportionality when animal interests are at stake? Third, do the rules

---

[11]ICRC, 'Customary International Law Database: Rule 44', available at: https://ihl-databases.icrc. org/customary-ihl/eng/docs/v1.

[12]Sassòli, 'Challenges faced by non-state armed groups' 2014, 172.

[13]AP I, 1977 (n. 5), art. 52(2).

[14]AP I, 1977 (n. 5), art. 51(5)(b) & art. 57.

on the conduct of hostilities adequately protect animals as 'inanimate objects'? If not, how could they be better treated?

With respect to the first question, it could be contended that the notion of 'objects' is limited to 'inanimate objects,' as exemplified by the list of objects cited in Article 52(3) of AP I, which mentions, among other things, places of worship, houses, schools, weapons, transport, fortifications, etc. Animals seem thus to be left unprotected.[15] It is true that, aside from a few multilateral treaties protecting endangered species, the welfare of animals has always been poorly regulated at the international level.[16] As noted previously, this is especially valid for IHL, which is, by tradition, eminently anthropocentric. That said, the context in which IHL conventions have been adopted has significantly changed over the last decades, especially since the Vietnam War. There is today a general rise in public awareness of the necessity to improve the protection of animals whose lives are threatened by modern warfare and, in particular, by the use of certain weapons, such as mines, cluster munitions, or by the destruction of oil installations. For obvious reasons, animals cannot be assimilated into the category of 'protected persons' under IHL, thereby benefiting from the protection offered by the status of 'combatant/prisoner of war' or of 'civilian.' Indeed, as rightly pointed out by Marco Roscini, 'if they were considered combatants, animals would have not only the rights, but also the obligations associated with this status (...).'[17] Clearly, animals are not able to respect these obligations, which, among other things, require the capacity to distinguish persons who participate in the hostilities from those who do not or to make proportionality calculations.[18] Moreover, the definition of 'civilians' mentioned in Article 50(1) of AP I only refers to 'persons.'[19] Thus, in order to avoid any gap in protection, animals cannot but fall within the category of objects as envisaged in IHL conventions. After all, this is consistent with the fact that, in most legal systems, animals have traditionally been considered as being 'moveable objects'. In light of this, animals could only be targeted in times of war when used for military purposes and when targeting them offers a definite military advantage. It should lastly be noted that, on a textual level, Article 54 of AP I grants a specific protection to objects that are 'indispensible for the survival of the civilian population.' This provision—which is located in Chapter III of AP I entitled 'Civilian objects'—includes livestock among these goods. This clearly confirms that certain animals are assimilated to objects by IHL.[20] However, such provision—which reflects IHL utilitarian approach that values animals for what it offers to human beings and which mainly aims at preventing starvation—mostly encompasses cattle, but not many other types of animals that also deserve protection.

---

[15]Roscini, 'Animals and the Law of Armed Conflict' 2017, 46.

[16]Peters, 'Global Animal Law: What It Is and Why We Need It' 2016, 13-16.

[17]Roscini, 'Animals and the Law of Armed Conflict' 2017, 44-45.

[18]Nowrot, 'Animals at War' 2015, 140.

[19]Roscini, 'Animals and the Law of Armed Conflict' 2017, 45.

[20]Ibid.

As to the second question, armed forces must make a proportionality calculation when launching an attack on civilian objects that might impact on animals. This calculation is a complex endeavour since it requires balancing incomparable elements: the military advantage of an attack, and the effect on civilian objects resulting from that attack. Such an exercise becomes increasingly difficult when animals may become collateral damage.[21] Indeed, it depends on the value attributed to animals. In most societies, such a value judgment is contingent upon what animals offer to humankind: working tool, food, clothing, etc. It is, however, increasingly accepted that animals should also acquire value in their own right and, as a consequence, that their interests should no longer be automatically subordinated to those of human ones. A greater intrinsic value is sometimes attributed to animals whose extinction is implicated. To further complicate matters, the status of animals varies widely from one culture to another and inevitably changes over time. Moreover, a second dimension must be a factor in the proportionality test: animal considerations must be weighed against human considerations (as opposed to military advantage).[22] For instance, when soldiers responsible for an attack are at risk, should these soldiers assume an increased risk in order to guarantee the protection of dogs or horses located around a military target? What should they do differently, if anything, if these animals are endangered species such as pandas or white rhinoceroses?

Without pretending to solve these issues here, three principles should guide belligerents in this regard. First, all measures should be taken to avoid any collateral damage to endangered species. This rule is grounded on the absolute necessity of sparing vulnerable animals from hostilities and other human activities that threaten them, as illustrated by the Convention on International Trade in Endangered Species of Wild Fauna and Flora,[23] which prohibits unpermitted trading. Second, all measures should be taken to avoid collateral damage to animals that are necessary for the survival of the civilian population. This second rule is grounded on Article 54 of AP I referred to above that grants special protection to these types of 'objects.' Third, the welfare of all other animals, including wild animals, should be duly taken into account when undertaking a proportionality assessment, as long as soldiers launching an attack could reasonably foresee damage that could be caused to these animals. Indeed, it would be unreasonable to impose upon belligerents a duty to limit animal casualties that are, for instance, so small that they are impossible to identify. In principle, when human interests conflict with animal interests, human life or suffering should prevail over those of animals.

The third question—the treatment of animals as objects—is reductive. In many jurisdictions around the world, animals are formally recognized as 'sentient beings.'

---

[21]For an extensive analysis of the concept of proportionality in the context of the protection of the environment, see generally Schmitt, 'Green War' 1997, 55-61.

[22]Ibid., 58-61.

[23]Convention on International Trade in Endangered Species of Wild Fauna and Flora, 3 March 1973, 993 UNTS 24. It should however be noted that the application of CITES during warfare remains unclear.

Of course, no state would be ready to create a new category of living beings that could limit their capacities to fight other belligerents, including armed groups. That said, in the same manner that human rights have progressively offered a series of protections to human beings in armed conflicts,[24] animal rights could progressively influence the way certain existing IHL principles are interpreted and applied to animals. Three examples illustrate this proposition. First, IHL prohibits the use of means and methods of warfare that cause superfluous injury or unnecessary suffering.[25] This prohibition only refers to the effect of weapons on combatants. Their impact on animals would never be taken into account so long as animals are regarded as 'inanimate objects.' Such an approach should, however, be reassessed if it is formally acknowledged that, for the purposes of IHL, animals, like any living beings, experience emotions, including pain and distress, which could be negatively affected by the use of these weapons. Second, as noted in the introduction, animals can often be used to perform military tasks, thereby making a contribution 'by their purpose or use' to military actions. When they offer a definite military advantage, these animals could, in principle, be 'destroyed', 'captured' or 'neutralized.'[26] Indisputably, this principle entails that, when animals previously categorised as military objectives lose the characteristics that allowed them to be so classified, they revert to being civilian objects until such time as they may again fulfil the qualification of military objective. Moreover, in the same manner that the fundamental principle of 'human dignity' prohibits the use of inhumane methods and means of warfare against combatants, a similar concept of 'animal dignity' could offer an avenue for expanding the safeguards offered to animals involved in hostilities, thereby preventing the use of indiscriminate techniques or unnecessary suffering.[27] Third, should animal be treated as sentient beings or be afforded legal personality, animal welfare should also carry heavier weight in the proportionality calculation than any inanimate objects in times of armed conflict.

5 The Protection of Individuals

The rules, which aim at protecting injured, sick and shipwrecked persons as well as prisoners of war and civilians, were not designed for animals. Three important exceptions must, however, be mentioned. First, Article 35 of GC I provides general protection to means of transport of wounded and sick or of medical equipment. When used for that purpose, animals should benefit from this safeguard. Second,

---

[24]See generally Doswald-Beck, *Human Rights in Times of Conflict and Terrorism* 2012.

[25]AP I, 1977 (n. 5), art. 35(2). For an extensive analysis of the application of such a prohibition to animals, see generally Roscini, 'Animals and the Law of Armed Conflict' 2017, 51-56.

[26]AP I, 1977 (n. 5), art. 52(2).

[27]On this basis, it could be argued that the use of animals during warfare must be prohibited at all times.

Article 60 of AP I confers protection to 'demilitarized' zones between belligerents.[28] This provision allows that an agreement on a demilitarized zone be tailored to each specific situation. It thus follows that the protection of areas of high global species diversity could be enhanced if belligerents were to agree to formally classifying them as 'demilitarized zones.' To this end, a Draft Convention on the Prohibition of Hostile Military Activities in Protected Areas has been developed following the 1990–1991 Gulf War in response to intensifying concerns about ecosystem damages during warfare.[29] To date, however, this Convention has not received the diplomatic support needed for its adoption. It is also worth mentioning here the importance of the 1972 Convention for the Protection of the World Cultural and Natural Heritage[30] in which States recognize the duty to identify and safeguard certain places that constitute part of the common heritage of humankind, including the habitat of threatened species of animals 'of outstanding universal value from the point of view of science or conservation.'[31] This Convention expressly states that the 'the outbreak or the threat or an armed conflict' is sufficient to place a property on the World Heritage in Danger list.[32] The third exception pertains to the laws governing occupation. Article 53 of GC IV prohibits the destruction by the occupying power of private and public properties, except in cases of absolute military necessity. This provision could provide minimum protection to certain animals when considered to be items of private or public property.

The acquisition by animals of the status of 'sentient beings' should also contribute to further development of the protection of animals when they fall into the hands of belligerents. Three main avenues could be explored in this regard. First, animals, in particular protected species, should receive adequate care when wounded during hostilities, and be evacuated as soon as possible from combat zones. The nature and the extent of such obligations should vary in functions of the capacities of the belligerents and the concrete situation on the battlefield. Second, animals involved in hostilities and apprehended should be granted basic 'humane' treatments tailored to their specific needs. For instance, they should never be killed without reason. They should receive food and adequate protection and, if they can survive by themselves, be released in areas far away from the battlefield. Of course, the furniture of such treatment should never impact on the entitlements offered by IHL to human beings, which must remain a priority. Third, the obligations of occupying powers vis-à-vis animals should be clarified in IACs and expanded to NIACs where the trafficking of expensive animal products often occurs. In this regard, the provisions allowing the

---

[28]United Nations Environment Programme (UNEP), Protecting the Environment During Armed Conflict. An Inventory and Analysis of International Law (Nairobi: UNEP 2009), 20.

[29]Convention on the Prohibition of Military or Any Other Hostile Use of Environmental Modification Techniques (ENMOD), 10 December 1976, 1108 UNTS 151.

[30]Convention for the Protection of the World Cultural and Natural Heritage, 16 November 1972, 1037 UNTS 151.

[31]Ibid., art. 2.

[32]Ibid., art. 11 (4).

occupying power to use occupied property without damaging or destroying it, as contained in the 1907 Hague Convention IV,[33] may offer guiding principles for dealing with similar situations in NIACs.[34]

As we can see, the protection of animals during warfare generates many complex questions. The international community should address these issues at a time when—as a matter of fact—animals are increasingly suffering from the extreme violence of modern warfare and when—as a matter of law—their status as 'sentient beings' has been acquiring wider recognition at national level.

References

Daskin, J. H., & Pringle, R. M. (2018, January 18). Warfare and wildlife declines in Africa's protected areas. *Nature, 553*, 328–332.

Doswald-Beck, L. (2012). *Human rights in times of conflict and terrorism.* Oxford: Oxford University Press.

Hanson, T., Brooks, T. M., Da Fonseca, G. A. B., Hoffmann, M., Lamoureux, J. F., Machlis, G., et al. (2009). Warfare in biodiversity hotspots. *Conservation Biology, 23*, 578–587.

Nowrot, K. (2015). Animals at War: The status of "animal soldiers" under international humanitarian law. *Historical Social Research, 40*, 128–150.

Peters, A. (2016). Global animal law: What it is and why we need it. *Transnational Environmental Law, 5*, 13–16.

Roscini, M. (2017). Animals and the law of armed conflict. *Israel Yearbook on Human Rights, 47*, 35–67.

Sassòli, M. (2014). Challenges faced by non-state armed groups as regards the respect of the law governing the conduct of hostilities. In E. Geppi (Ed.), *Conduct of hostilities: The practice, the law and the future* (pp. 171–178). San Remo: International Institute of Humanitarian Law.

Schmitt, M. N. (1997). Green War: An assessment of the environmental law of armed conflict. *Yale Law Journal, 22*, 1–110.

Jérôme de Hemptinne is teaching international humanitarian law at the Universities of Louvain and Lille and at the Institut d'études politiques (Paris). He is currently a researcher at the University of Geneva. His research focuses on the qualification of armed conflicts, modes of liability in international criminal law and institutional aspects of international criminal courts and tribunals. He worked for nearly two decades at the International Criminal Tribunal for the Former Yugoslavia, at the Office of the UN Legal Counsel in New York and at the Special Tribunal for Lebanon. He is a member of the editorial committee of the *Journal of International Criminal Justice*.

---

[33]Convention (IV) respecting the Laws and Customs of War on Land and its annex: Regulations concerning the Laws and Customs of War on Land, 18 October 1907, 205 CTS 277.

[34]Convention for the Protection of the World Cultural and Natural Heritage (n. 30), 19.